Photocatalysts for Energy and Environmental Sustainability

Online at: https://doi.org/10.1088/978-0-7503-5697-8

Photocatalysts for Energy and Environmental Sustainability

Edited by
Vijay B Pawade
Department of Applied Physics, Laxminarayan Innovation Technological University, Nagpur 440033, India

Bharat A Bhanvase
Department of Chemical Engineering, Laxminarayan Innovation Technological University, Nagpur 440033, India

IOP Publishing, Bristol, UK

ISBN 978-0-7503-5697-8 (ebook)
ISBN 978-0-7503-5693-0 (print)
ISBN 978-0-7503-5700-5 (myPrint)
ISBN 978-0-7503-5694-7 (mobi)

DOI 10.1088/978-0-7503-5697-8

Version: 20240301

IOP ebooks

British Library Cataloguing-in-Publication Data: A catalogue record for this book is available from the British Library.

Published by IOP Publishing, wholly owned by The Institute of Physics, London

IOP Publishing, No.2 The Distillery, Glassfields, Avon Street, Bristol, BS2 0GR, UK

US Office: IOP Publishing, Inc., 190 North Independence Mall West, Suite 601, Philadelphia, PA 19106, USA

Contents

Preface ix

Editor biographies x

List of contributors xii

1 An introduction to photocatalysts and their applications 1-1
Vijay B Pawade and Bharat A Bhanvase

1.1 Introduction 1-1
1.2 The principles and mechanism of photocatalysis 1-2
1.2.1 Types of photocatalyst 1-4
1.2.2 Single-atom photocatalysts 1-5
1.2.3 Quantum-dot-based photocatalysts 1-5
1.2.4 Perovskite-based photocatalysts 1-6
1.2.5 S-scheme photocatalysts 1-6
1.2.6 rGO-based composite photocatalysts 1-7
1.2.7 Semiconductor photocatalysts 1-8
1.3 Synthesis methods 1-8
1.3.1 Chemical methods 1-8
1.3.2 Green synthesis 1-9
1.4 The reusability and stability of photocatalysts 1-10
1.5 Factors affecting photocatalytic activity 1-10
1.5.1 The bandgap 1-10
1.5.2 Particle size 1-11
1.5.3 Doping 1-12
1.6 Strategies for boosting photocatalytic efficiency 1-12
1.6.1 Spatial separation of excitons 1-12
1.7 Applications 1-16
1.7.1 Energy sustainability 1-16
1.7.2 Environmental sustainability 1-17
1.8 Conclusions and future prospects 1-18
References 1-18

2 Rare-earth-doped metal oxide photocatalysts for wastewater treatment 2-1
G A Suganya Josephine, S Rubesh Ashok Kumar and D Vasvini Mary

2.1 Introduction 2-1

2.2 Semiconductor metal oxides 2-2
2.3 Rare-earth elements 2-4
2.4 The preparation of rare-earth-doped metal oxides 2-4
2.5 The photocatalytic removal of contaminants by rare-earth–metal oxides 2-6
2.6 The photocatalytic removal of antibiotics (endocrine disrupters) by rare-earth–metal oxides 2-8
2.7 The photocatalytic removal of organic dye effluent by rare-earth-doped metal oxides 2-9
2.8 The photocatalytic mechanism of rare-earth-based photocatalysts 2-10
2.9 Conclusions 2-11
References and further reading 2-12

3 Composite nanostructures based on metal oxide/upconversion phosphors with improved photocatalytic activity 3-1
Kamila Zhumanova, Darya Goponenko and Timur Sh Atabaev

3.1 Introduction 3-1
3.1.1 Metal oxide semiconductors 3-1
3.1.2 Upconversion materials 3-4
3.2 Composite nanostructures based on metal oxide/upconversion phosphors 3-6
3.2.1 Upconversion-capable photocatalytic materials for water treatment 3-7
3.2.2 Upconversion-capable photocatalytic materials for hydrogen generation 3-11
3.3 Conclusions and future prospects 3-14
Acknowledgments 3-15
References and further reading 3-15

4 Nanocomposites and their applications in photocatalytic degradation processes 4-1
Shubham Bonde, Gauri Kallawar, Bharat A Bhanvase and Bhaskar Sathe

4.1 Introduction 4-1
4.2 Nanocomposite materials 4-2
4.2.1 Types of nanocomposite materials 4-3
4.2.2 Methods of nanocomposite synthesis 4-11
4.2.3 Properties of nanocomposite photocatalysts 4-24

4.3 The fundamentals of photocatalysis 4-26
4.3.1 Photocatalysis 4-26
4.3.2 What are photocatalysts? 4-28
4.3.3 Desirable properties of photocatalysts 4-29
4.3.4 How do photocatalysts work? 4-31
4.3.5 Mechanisms of photocatalysis 4-32
4.3.6 Factors influencing photocatalytic activity and efficiency 4-32
4.4 Nanocomposites for photocatalytic degradation 4-34
4.5 Challenges and future perspectives 4-35
4.6 Conclusions 4-37
References 4-38

5 Photocatalysts for hydrogen evolution 5-1
Somnath C Dhawale and Bhaskar R Sathe

5.1 Introduction 5-1
5.2 The fundamentals of photocatalytic water splitting 5-6
5.3 Quantum efficiency and solar-to-H_2 efficiency 5-8
5.4 Overall water splitting (hydrogen evolution/oxygen evolution) 5-9
5.5 The basic concepts of photocatalysis 5-10
5.6 Elements of photocatalytic hydrogen generation physics and chemistry 5-12
5.7 Systems for producing hydrogen via photocatalysis 5-13
5.8 Current advances in photocatalytic water splitting 5-15
5.9 Overall water splitting using photocatalysts 5-19
5.9.1 Photocatalyst systems with a single photocatalyst 5-20
5.10 Hybrid photosynthesis system: combining natural and artificial approaches 5-20
5.11 Conclusions and future perspectives 5-21
References and further reading 5-22

6 Photocatalysis for organic degradation using perovskite materials 6-1
Phuong Hoang Nguyen, Thi Minh Cao and Viet Van Pham

6.1 Introduction 6-1
6.2 An overview of perovskite materials 6-3
6.2.1 An introduction to perovskites 6-3
6.2.2 The properties of perovskites 6-6
6.3 Methods used to synthesize perovskite materials 6-10

6.3.1 Wet chemical methods 6-10
6.3.2 Biotemplate-supported synthesis 6-12
6.4 Perovskite photocatalysts for organic degradation 6-14
6.4.1 Photocatalysis 6-15
6.4.2 Fenton and photo-Fenton catalysis 6-18
6.4.3 PMS activation 6-19
6.5 Conclusions 6-21
References 6-22

7 Emerging trends and future prospects in photocatalysis-based environmental remediation and hydrogen production 7-1
İbrahim Hakkı Karakaş and Zeynep Karcıoğlu Karakaş

7.1 Introduction 7-1
7.1.1 Emerging trends and future prospects in photocatalysis 7-1
7.2 Scalability challenges for photocatalytic applications 7-4
7.3 Scalability challenges for photocatalytic H_2 production applications 7-7
7.4 Advances in photocatalytic materials and innovative synthesis approaches 7-9
7.5 The future evolution of light sources for photocatalytic processes 7-13
7.5.1 Plasmonic photocatalysis 7-16
7.5.2 A comparison of visible-light photocatalysis and plasmonic photocatalysis 7-17
7.6 Conclusions 7-18
References 7-19

Preface

In recent years, research into energy and environmental sustainability has received more importance in interdisciplinary areas of science and engineering due to the vast increase in industrial globalization. So, many research organizations and academic institutes are promoting these frontier areas as a way of developing highly efficient, environmentally friendly technology to achieve sustainability goals. Thus, photocatalysis is an emerging, simple, and low-cost technique that has the potential to resolve issues related to hydrogen generation and the photocatalytic degradation of pollutants under sunlight illumination. This textbook summarizes the fundamental mechanisms, properties, and applications of different types of photocatalysts. It contains seven chapters that cover the current progress and future scope of new and advanced photocatalytic materials, written by well-known authors in these fields. Therefore, this textbook is designed to be of benefit in undergraduate as well as postgraduate courses in science and technology. As per the global scope of environmental research, this book can provide an ideal platform for the reader to understand the concepts presented in a more systematic way, increasing their interest in the content of the book. So, we thank all the contributing authors for their efforts to enhance the depth of this book and their expertise in making this textbook attractive among the other books. We also thank IOP Publishing for introducing this textbook on new and advanced photocatalytic materials and their sustainable approach for the betterment of mankind. We sincerely hope this book can ultimately make a significant contribution to research and development activities in the field of photocatalysis.

Editor biographies

Vijay B Pawade

Dr Vijay B Pawade is an assistant professor (Sr.Gr) in the Department of Applied Physics at the Laxminarayan Innovation Technological University, Nagpur, India. His research focuses on rare-earth-doped oxide materials and their applications in light-emitting diodes (LEDs), solar cell devices and photocatalytic processes. He has published 45 research papers in respected international peer-reviewed journals and acts as a reviewer for journals published by Elsevier, Springer, Wiley, Taylor and Francis, the Royal Society of Chemistry, and the American Chemical Society. He has contributed 12 book chapters on different themes such as nanomaterial synthesis and characterization, the applications of nanomaterials in energy conversion and storage devices, quantum dots (QDs), the spectroscopy of lanthanides, etc. He is the author of books titled Phosphor for Energy Saving and Conversion Technology (***CRC Press—Taylor and Francis***), Optical Properties of Phosphate and Pyrophosphate Compounds, and Lanthanide-Doped Aluminate Phosphors (***Woodhead Publishing—Elsevier***). He has edited five books on Nanomaterials for Green Energy (***Elsevier***), Spectroscopy of Lanthanide-Doped Oxide Materials (***Woodhead Publishing—Elsevier***), Multifunctional Nanostructured Metal Oxides for Energy Harvesting and Storage Devices (***CRC Press—Taylor and Francis***), Handbook of Nanomaterials for Wastewater Treatment (***Elsevier***), Nanoscale Compound Semiconductors and their Optoelectronics Applications (***Woodhead Publishing—Elsevier***), Phosphor Handbook: Process, Properties and Applications (***Woodhead Publishing—Elsevier***).

Bharat A Bhanvase

Dr Bharat A Bhanvase is currently working as professor and head of the Chemical Engineering Department at the Laxminarayan Innovation Technological University, Nagpur, India. His research interests are focused on wastewater treatment, cavitation-based synthesis of nanomaterials and nanocomposites, solid waste processing, process intensification, microfluidics, nanofluids, etc. He obtained his PhD in Chemical Engineering from the University of Pune. He has published 103 articles in international journals, and four in national journals; he has presented 17 papers in international conferences and 12 in national conferences. He has written 55 book chapters in internationally renowned books, nine edited books, and one authored book. He has obtained four Indian patents and applied for six Indian patents. He received IIChE Awards in 2021 (the Chemical Weekly Award, the IIChE NRC Award, and the Kuloor Memorial Award) for the best paper published in the 'Indian Chemical Engineer' in its 2020 issues. He is a fellow of the Maharashtra Academy of Sciences (MASc) and a fellow of IIChE. He was the recipient of the Best Scientist Award from Rashtrasant Tukadoji Maharaj Nagpur University in 2017. He also received a Young Scientists (Award) start-up research grant from the Science and Engineering Research Board, New Delhi (India) in 2015. He has guided 28 M.Tech. students and one PhD student; two M.Tech. and three PhD students are currently working with him.

List of contributors

Timur Sh Atabaev
Department of Chemistry, Nazarbayev University, Astana 010000, Kazakhstan

B A Bhanvase
Department of Chemical Engineering, Laxminarayan Innovation Technological University, Nagpur 440033, India

Shubham Bonde
Department of Chemical Technology, Laxminarayan Innovation Technological University, Nagpur 440033, India

Thi Minh Cao
HUTECH University, 475A Dien Bien Phu Street, Binh Thanh District, Ho Chi Minh City, Viet Nam

Somnath C Dhawale
Department of Chemistry, Dr Babasaheb Ambedkar Marathwada University, Chatrapati Sambhajinagar 431004, MH, India

Darya Goponenko
Department of Chemistry, Nazarbayev University, Astana 010000, Kazakhstan

G A Suganya Josephine
Department of Chemistry, Center for Nanotechnology Research, Aarupadai Veedu Institute of Technology—Vinayaka Mission Research Foundation, Rajiv Gandhi Salai, Paiyanoor, Kanchipuram 603104, India

Gauri Kallawar
Department of Chemical Technology, Dr Babasaheb Ambedkar Marathwada University, Aurangabad 431004, MS, India

İbrahim Hakkı Karakaş
Department of Food Engineering, Bayburt University, Bayburt, Turkey

Zeynep Karcıoğlu Karakaş
Department of Environmental Engineering Atatürk University, Erzurum, Turkey

S Rubesh Ashok Kumar
Department of Chemistry, Center for Nanotechnology Research, Aarupadai Veedu Institute of Technology—Vinayaka Mission Research Foundation, Rajiv Gandhi Salai, Paiyanoor, Kanchipuram 603104, India

D Vasvini Mary
Department of Chemistry, Center for Nanotechnology Research, Aarupadai Veedu Institute of Technology—Vinayaka Mission Research Foundation, Rajiv Gandhi Salai, Paiyanoor, Kanchipuram 603104, India

Phuong Hoang Nguyen
HUTECH University, 475A Dien Bien Phu Street, Binh Thanh District, Ho Chi Minh City, Viet Nam

V B Pawade
Department of Applied Physics, Laxminarayan Innovation Technological University, Nagpur 440033, India

Viet Van Pham
HUTECH University, 475A Dien Bien Phu Street, Binh Thanh District, Ho Chi Minh City, Viet Nam

Bhaskar R Sathe
Department of Chemistry, Dr Babasaheb Ambedkar Marathwada University, Chatrapati Sambhajinagar 431004, MH, India

Kamila Zhumanova
Department of Chemistry, Nazarbayev University, Astana 010000, Kazakhstan

Chapter 1

An introduction to photocatalysts and their applications

Vijay B Pawade and Bharat A Bhanvase

This chapter introduces the fundamentals of photocatalysts and their role in the development of sustainable technologies. It explains the basic principles, mechanisms, and workings of photocatalysts for wastewater treatment and hydrogen generation. It also explores the different types of photocatalysts, including their characteristics and features at both the nanoscale and the microscale. The methods of synthesis and the importance of green synthesis compared to other conventional routes are discussed in detail. In addition, this chapter discusses some other important parameters reported in the research literature, such as the reusability and stability of photocatalysts, factors affecting the photocatalytic performance of photocatalysts, and the need for new and advanced strategies to improve the photocatalytic efficiency of photocatalysts for the production of energy and the development of environmentally friendly technology.

1.1 Introduction

In the last few decades, a major worldwide focus has been placed on the development of sustainable technology to protect the environment and maintain the harmony of nature on our mother planet. It is our social responsibility to protect and balance the environment through the development of new and advanced sustainable technologies. Nowadays, there is tremendous growth and strong competition everywhere due to the supply of, and demand for products in the global market due to the vast increase in populations in underdeveloped countries. Thus, to meet the need for low-cost products and the need to recycle cheaper raw materials, many new small- and large-scale industries have been set up to fulfill the global demands for materials and related products. During the recycling of technologically outdated products and devices, many kinds of toxic gases and heavy elements are

doi:10.1088/978-0-7503-5697-8ch1

produced and dispersed into the air, water, and soil; these not only affect the quality of the air, water, and soil but also increase their toxicity levels. Thus, among these three sources of pollution, the prevention of air and water pollution are the top priorities, as they impact all living things on our planet. They also have a major impact. In terms of the social and economic development of nations, further increases in pollutant levels in fresh air and water increase the potential risks to people's health and cause many health issues related to respiratory disorders, dermatitis, asthma, mutagenicity, cancer, etc [1]. Photocatalysis is a more promising and sustainable way to resolve to resolve such global environmental issues related to air and water pollution than other techniques. Because it can effectively convert and utilize solar energy, it has received more attention in recent years for its prospective use in photocatalytic wastewater treatment and advanced water splitting processes for the production of H_2, which is considered to be a clean and environmentally friendly source of energy [2, 3]. Basically, a photocatalyst is a material involved in specific chemical reactions that take place under exposure to light radiation, in which it converts the solar energy into other useful forms. For the photocatalytic process, sunlight is the most prominent inexhaustible and clean source of driving energy that leads to slow reaction conditions, high energy of the active species, and a deep oxidation effect during the photocatalytic reaction. This reaction can be caused by the absorption of sunlight in different regions of the spectrum, such as UV, visible light, and infrared radiation; the specific region involved usually depends on the photocatalyst material [4–7]. Thus, the photocatalytic process has the ability to resolve the problems related to the environment and energy without the utilization of excessive fossil fuels. Photocatalysts are capable and operate well under natural sunlight, but more effort is needed for the development of highly efficient visible-light-driven photocatalyst materials [8]. Khan *et al* [9] reported that research and development into the photocatalytic process exhibits a broad scope for widespread application in the near future. However, a few parameters of photocatalytic materials, such as their efficiency, thermal stability, purity, environmental compatibility, and low efficiency in photocatalytic reactors, are major hurdles that restrict their application at scale. In recent years, inorganic bandgap semiconductors such as ZnO, CdS, ZnS, ZnSe, CdSe, ZnTe, etc. have been widely studied for the photocatalytic process due to their unique chemical properties and high stability [10–13]. Among these, zinc oxide (ZnO) is the most favorable, environmentally friendly, and economically viable for large-scale wastewater treatment. In the following section, we discuss the details, principles, and working mechanisms of the photocatalytic materials and their types that are proposed for the advanced water treatment and hydrogen generation processes.

1.2 The principles and mechanism of photocatalysis

The photocatalytic process is based on the absorption of light by the photocatalyst, for which metal-oxide semiconductors are preferable because of their suitability for the formation of electron–hole pair creation in the conduction band (CB) and the valence band (VB) [14]. Thus, during the absorption of light, electrons in the VB are

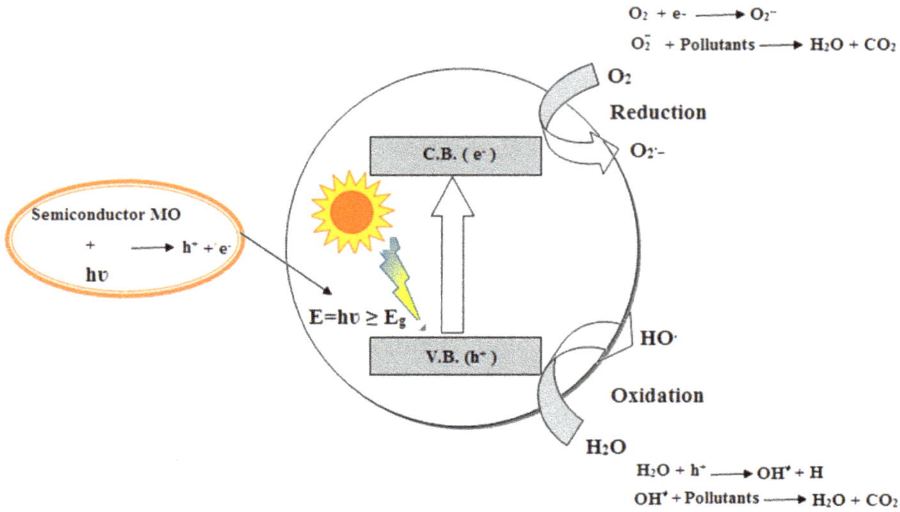

Figure 1.1. The photocatalytic mechanism under solar illumination.

excited into the CB where they form electron (e^-) and hole (h^+) pairs. There are two photochemical reactions that involve the photoinduced electrons and holes, which are continuously generated. A schematic of the photocatalytic process and its mechanism of pollutant degradation under solar illumination are shown in figure 1.1. In general, photocatalytic materials play a major role in initiating the oxidation and reduction reactions in the presence of solar energy. The following steps are followed during the photocatalytic process:

Step **I. The** generation of hole/electron pairs

Step **II.** The separation of charge carriers and their diffusion towards the electrode surface

Step **III.** Photooxidation and -reduction reactions take place at the surface of the photocatalyst.

Here, photoinduced holes and electrons react with O_2 and H_2O on the photocatalyst's surface, which leads to the formation of O^{-2} and $OH^{\bullet}$ radicals. These radicals have strong redox potentials, and hence, when they react with pollutants, photodegradation takes place. A possible photocatalysis reaction is shown in figure 1.1.

When active species are adsorbed by the photocatalyst's surface, the electron transfer process becomes more prominent [15]. Further, during the water cleaning process, oxygen acts as a common electron acceptor. When photogenerated electrons react with oxygen, they reduce to $O^{-}{}_2$ and can be transformed into various oxygen-activated species, such as $HO^{\bullet}$, H_2O_2, $HO_2{}^{\bullet}$ and HO_2 anions [16, 17], which involve the oxidation of the electron donor [18], while the generated holes can oxidize the electron donor. At the same time, reactive oxidizing species and free carriers react with absorbed surface impurities, and the degradation of pollutants takes place. The efficiency of the photocatalysis process depends on the ability of the photocatalyst to produce a large number of holes and electrons, which results in the production of reactive free radicals. A shift in the light absorption range of the photocatalyst into the visible spectral range helps to generate a large number of electron–hole pairs, thereby improving the degradation response of the photocatalyst [19]. Hence, the absorption range plays a key role for highly active photocatalysts.

1.2.1 Types of photocatalyst

As compared to conventional water treatment processes, advanced oxidation processes are assigned great importance due to their stronger oxidation capabilities, faster reaction times, and production of smaller amounts of secondary pollutants. These processes are generally categorized into homogeneous and heterogeneous processes, depending on the type of reaction medium. Further, they can be classified into energy- and non-energy-related categories [20]. The abovementioned heterogeneous photocatalytic oxidation technique is widely accepted for the degradation of pollutants in wastewater [21]. This technique has some advantages, such as flexibility, simplicity, low cost, the use of an environmentally friendly catalyst, and high photocatalytic efficiency. In the recent years, many new types of photocatalysts have been proposed and used for the removal of organic pollutants from wastewater [22, 23], Some of these are discussed below.

1.2.1.1 Homojunction semiconductor photocatalysts

Homojunction semiconductor photocatalysts are synthesized by incorporating semiconductor interfaces that have compatible bandgap energies and chemical compositions and specific dimensions [24]. They also possess particular physical, electrical, and optical properties and exhibit superior photocatalytic activities for the photodegradation of waterborne organic pollutants [25]. Nanoscale photocatalytic devices have been fabricated by using homojunction photocatalysts and are applicable in various disciplines [26, 27]. Further, homojunction photocatalysts help to improve photocatalytic efficiency in the production of hydrogen via the water splitting process [28]. But the use of semiconductors for photocatalytic applications has some limitations.

1.2.1.2 Hetrojunction semiconductor photocatalysts

Heterogeneous semiconductor photocatalysis is another type of advanced oxidation process that has received great attention due to its prospective use in resolving energy and environmental issues by, for example, generating hydrogen through the water splitting process and degrading organic pollutants through redox reactions [29]. Semiconductor heterojunctions are constructed by combining two semiconductors, and they have been demonstrated to be one of the most efficient ways to spatially separate photoexcited electron–hole (e^-/h^+) pairs [30, 31]. When a heterojunction photocatalyst is illuminated by a light source, photoexcited charge carriers are forced to move between the two semiconductors, building up an electric field and hence inhibiting the recombination of the charge carriers. The formation of a built-in electric field at the semiconductor heterojunction interface and the transfer rate of a photoexcited charge carrier depend on the semiconductivity of the materials, the work function, and the ratio of the CB to VB potentials of the semiconductors.

1.2.2 Single-atom photocatalysts

Single-atom photocatalysts (SACs) are considered to be low-cost, high-efficiency photocatalysts and have been assigned more importance in the field of catalysis [32]. Qiao *et al* [33] were the first to report the concept of 'single-atom catalysis' [34]. SACs have been synthesized by loading a single metal atom onto a suitable support; further electrons are exchanged with the support to form single-atom active sites, enhancing the photocatalytic performance of the material [35]. Due to continuous research into, and development of the preparation of SACs, many preparation techniques have come into existence. Thus, the flexible pairing of metal centers and charge carriers facilitates the preparation of environmentally friendly and sustainable single-atom photocatalysts with high catalytic efficiency [36, 37]. SACs are also used to produce both homogeneous and heterogeneous catalysts [38, 39]. The absorption range and charge separation efficiency of SACs are high [40]. As a result of these characteristics, SACs are emerging materials in the photocatalytic field for the evolution of photocatalytic H_2 and the removal of toxic contaminants from wastewater [41, 42].

1.2.3 Quantum-dot-based photocatalysts

Quantum dot (QD)-based composite catalysts are considered to be promising candidates for resolving issues related to energy and environmental sustainability. QDs are zero-dimensional spherical nanoparticles, and their physical dimension is smaller than the exciton's Bohr radius [43, 44]. Colloidal semiconductor nanocrystal QDs 2–10 nm in size may contain 10–50 atoms within their volume [45]. Recently, Kandi *et al* [46] discussed the scope and advantages of quantum dots in the photocatalytic hydrogen production process. There are some characteristics of QDs that make them suitable for enhanced H_2 production compared to other types of nanostructured materials with superior properties; these characteristics play a significant role in enhancing photocatalytic activity. Some of the important properties of QDs are given below:

(i) The capability to absorb light in the visible spectral range
(ii) A better multiple-exciton generation rate under solar illumination due to the quantum confinement effect.
(iii) Better charge transport and separation characteristics.
(iv) Size-dependent tuneable optical properties.
(v) Their visible-light absorption edge can be enhanced by doping them with wide-bandgap semiconductors.

At present, many research groups are working on the development of highly efficient QDs based on hybrid systems that have the above characteristics and properties and are applicable for effective photocatalysis.

1.2.4 Perovskite-based photocatalysts

Perovskite-structured materials represented by the chemical formula ABX_3 belong to a ternary family of crystalline structures in which the A-site contains metal cations, rare earth ions, or alkaline earth metal ions with larger ionic radii and the B-site contains transition-metal ions with smaller ionic radii, while X indicates the oxygen atoms available in the host structure. Perovskite-structured materials exhibit many interesting properties; further, perovskite nanomaterials show excellent photocatalytic efficiency due to their characteristics such as superior chemical and thermal stability, nontoxicity, cost-effectiveness, tuneable properties, adjustable bandgap, large charge carrier lifetime, etc [47]. The shape- and size-dependent properties of a perovskite nanostructure depend on the method of synthesis and its structural characteristics. Today, perovskite nanoparticles have the potential to be used in a variety of applications, such as chemical sensing, catalysis, water splitting, and the photodegradation of organic pollutants. But single-component perovskite materials have a broader bandgap, and hence recombination of the charge carrier takes place much faster, which restricts their performance in visible-light-driven photocatalysis. For effective photocatalysis under solar illumination, strong absorption near 520 nm is needed. Further challenges still remain, such as resolving the problems of the separation and recycling of perovskite materials in treated water. To overcome these challenges, further research activity and strategies, such as modification of their surface, doping with metal ions, coupling with metal nanoparticles, the synthesis of nanocomposites, etc. [48], are required to improve the photocatalytic performance of these materials.

1.2.5 S-scheme photocatalysts

S-scheme heterojunction photocatalysts exhibit characteristics such as a superior light absorption ability, a high charge carrier separation efficiency, a strong redox potential, and a diverse range of both inorganic and organic semiconductors. Considering the related advantages and disadvantages of conventional heterojunction photocatalysts, the step-scheme (S-scheme) is a novel semiconductor catalyst that fulfills the current need for efficient photocatalysts [49]. Here, the heterojunction is formed by contact between two different semiconductors, which helps to increase the absorption band edge of the semiconductors, further improving the separation and migration rate of the photogenerated carriers [50]. As a result of the close contact between the different semiconductors, the electrons from the reduced semiconductor spontaneously migrate toward the oxidized semiconductor and build an electric field that is directed toward the oxidized semiconductor. At the same time, the holes in the VB of the reduced semiconductor and the electrons in the CB of the oxidized semiconductor combine, which results in the accumulation of a greater number of negatively charged carriers (i.e. electrons) in the CB of the reduced semiconductor and a greater number of positively charged carriers (i.e. holes) in the VB of the oxidized semiconductor. Thus, the S-scheme heterojunction photocatalyst shows strong redox capacity [51] and has a wide range of potential applications. S-scheme heterojunction photocatalysts can be categorized as inorganic–inorganic [52], inorganic–organic [53], or organic-organic composites [54]. The inorganic–inorganic types of S-scheme heterojunction photocatalysts are of great interest for the photocatalysis process.

1.2.6 rGO-based composite photocatalysts

Graphene is a zero-bandgap material, which restricts its applications, particularly in the electronics field. Therefore, doping graphene with heteroatoms can form a number of localized energy levels in its bandgap, and hence it can exhibit tuneable properties that make it responsive in visible light. However, there are some restrictions on the use of GO, such as its toxicity and corrosiveness, the explosive nature of the reducing agents, etc [55]. However, reduced graphene oxide (rGO) also exhibits excellent properties such as tunable electrical properties, transparency, and the ability to integrate with various photoactive surfaces to enhance their efficiencies. Further, it has good electrical conductivity and a large surface area and is a better substitute for pure graphene which can be synthesized at low production costs using a simple reduction process [56]. Thus, it makes rGO a desirable candidate for solar-driven photocatalytic applications. Coupling rGO with oxide materials promotes electron separation, boosting their photodriven activity in the visible spectral region and promoting the degradation of some harmful dyes. Thus, 0D, 1D, and 2D nanostructured semiconductors coupled with rGO nanocomposites play an important role in improving photocatalytic activity, hydrogen generation, nitroaromatic reduction, etc. In the case of semiconductor composites, coupling the semiconductor with rGO helps to separate the photogenerated charge carriers at the catalyst/RGO interface. Here, the nature of the semiconductor/rGO interface and defects in the rGO play an important role in enhancing the photocatalytic activity. Recently, Witjaksono *et al* synthesized and reported visible-light-driven N-doped rGO with reduced bandgap energy, i.e. from 3.4 to 2.2 eV [57]. This material reduced the electron–hole pair recombination rate by exhibiting the characteristics and features of a visible-light-driven photocatalyst [58]. Similarly, ferrite-based rGO nanostructures are magnetic materials and have good absorption in the visible spectral range. Further, they are strongly responsive to applied magnetic fields and can be readily recovered using conventional magnetic bars [59]. Today, rGO is one of the benchmark materials for improving the performance of some advanced materials used in the development of sustainable technology [59]. Recyclability and recovery are also important aspects of the development of new photocatalyst materials. Researchers have made many efforts to resolve current issues and future challenges in these fields.

1.2.7 Semiconductor photocatalysts

Semiconductor photocatalysts contain different materials such as metal oxides, nitrides, or sulfides (e.g. TiO_2 and MoS_2) [60, 61] as well as metal-free semiconductors such as C_3N_4. Other materials, such as copper, gold, and silver metal nanoparticles, exhibit strong localized surface plasmon resonance (LSPR) properties under visible-light irradiation. These nanoparticles were assigned great importance at the beginning of 21st century, but they have higher costs, which restricts their wider industrial scope. However, semiconductor photocatalysts are comparatively lower in cost and have been a topic of research for more than 50 years, but they suffer from the issue of a low absorption band in visible light and hence lower degradation efficiency. Based on some characteristics, features, and innovative approaches, they are considered promising materials for use in industrial applications at enhanced photocatalytic efficiencies. Semiconducting nanomaterial-based photocatalysts such as ZnO, CdS, ZnS, ZnSe, CdSe, ZnTe, etc. have been studied many times in the last few decades, and they are widely accepted due to their unique properties and good stability, which also promote strong redox reactions [10, 11]. ZnO is the most favorable, environmentally friendly, and economical catalyst for the large-scale treatment of wastewater due to its direct bandgap energy, which is of the order of 3.37 eV. It also seems to have an excellent degradation response under UV light illumination. The available wavelength spectrum of solar radiation contains only 4% of UV light but 43% of the visible-light component. To shift the response of ZnO under visible light, there is a need to alter the optical properties of ZnO by adding a narrow energy gap semiconductor, which improves its absorption capacity in visible light and also reduces the e^-/h^+ recombination rate [62]. ZnSe/ZnO is a well-known example of a composite catalyst, in which ZnO is combined with ZnSe, which acts as a narrower-bandgap semiconductor (2.67 eV). The bandgap of ZnSe is well aligned with that of ZnO, hence, it improves the photocatalytic degradation efficiency of ZnO [63]. However, much more investigation is still needed to explore the future prospects of advanced semiconductor materials to fully meet the needs of energy and environmental sustainability.

1.3 Synthesis methods

1.3.1 Chemical methods

In recent years, many conventional methods have been used to prepare nanostructures of different dimensions [64]. Thus, the top-down and bottom-up approaches are popular methods for the preparation of oxides and other forms of materials. The bottom-up approach is well suited for the fabrication of defect-free nanostructured materials of specific shapes and sizes. Some of the techniques, such as combustion, the hydrothermal method, the solvothermal method, the sonochemical method, the sol–gel method, etc. fall into the category of bottom-up approaches, as shown in figure 1.2. These are well-known methods used in materials synthesis. Among these techniques, great importance it attached to the green synthesis route, which involves the use of nonacid mediums as well as nontoxic materials for the

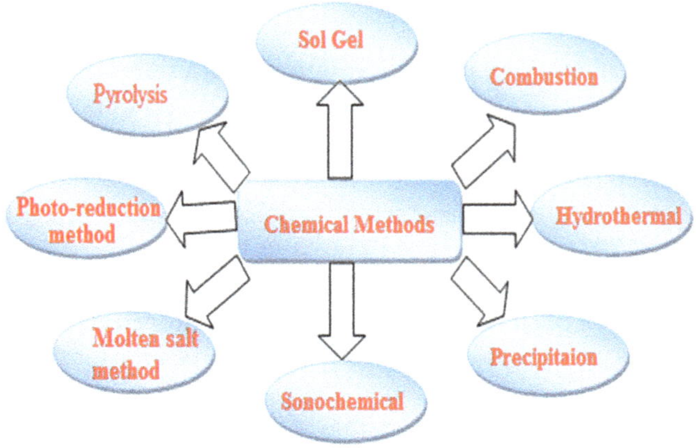

Figure 1.2. Chemical methods used for the synthesis of materials.

preparation of materials [64, 65]. Thus, the chemical approach is well suited for the preparation of inorganic nanostructured photocatalytic materials because of its ability to produce materials in the desired shape and size, and the methods used in the bottom-up approach are also cheaper than those of the top-down approach. But there is always a need to take care in the selection of nontoxic raw materials and the formation of the final pure-phase product while avoiding the presence of the acid medium, impurities, etc. High-temperature solid-state diffusion is also a well-known technique that corresponds to the top-down approach used to prepare perovskite oxide nanostructures, but this technique leads to the formation of an impure phase that contains inhomogeneous perovskite oxide nanomaterials due to repeatedly grinding, crushing, and preheating them before calcination. This leaves some defects on the surface of the nanostructured materials, and the presence of such defects may affect the properties of the nanostructured materials. In contrast, using the sol–gel method, the coprecipitation method, or the combustion method, it is possible to obtain pure perovskite oxide nanomaterials with a high surface area and an ideal nanostructure size. These are the simplest and most efficient techniques with which perovskite oxide materials are synthesized. Hydrothermal methods are also well-known synthesis techniques that allow the nanostructure's shape and size to be controlled. However, these techniques require precursors that readily mix well in aqueous solution at high temperature and constant pressure [66]. Therefore, the hydrothermal method is an important branch of inorganic synthesis that depends on the solubility of material in hot water under high pressure [67]. During the hydrothermal reaction, parameters such as the type of solvent, temperature, and time of reaction play an important role; in such reactions, nucleation and grain mechanisms form nanosized crystallites.

1.3.2 Green synthesis

Today, green synthesis is an environment-friendly technique for the preparation of nontoxic metal-oxide nanoparticles and has received great interest in the field of nanotechnology. During the synthesis of nanomaterials by chemical methods, some toxic gases are liberated during chemical reactions, and the harmful chemical species present are also adsorbed on the surfaces of nanoparticles. Thus, considering this drawback, green synthesis is an emerging technique for the production of NPs. Further, this method is clean, safe, and a cost-effective way to deploy environmentally friendly processes [68]. It is possible to synthesize nanoparticles of different shapes and sizes using this technique. The use of different fuels or reducing agents also plays an important role in obtaining the pure phase and the desired nanostructure morphology [69].

1.4 The reusability and stability of photocatalysts

Efficiency and reusability are two basic parameters that play an important role in the practical use of photocatalysts. They can be recycled at least five times, and the photocatalytic materials must remain stable during this process. The recovery of the materials can be carried out through the centrifugation technique after each cycle. The separated materials are then rinsed more than two times with deionized water. Later, the recovered materials are reused in the photocatalytic dye degradation process. Metal-oxide nanostructures such as acid protease functionalized silver nanoparticles (APTs–AgNPs) exhibit excellent catalytic performance that removes up to 95% of methylene blue (MB) dye from wastewater, and the reuse of the materials does not significantly alter their efficiency after each run. Hence, they can be reused more often with a minimal reduction in their efficiencies. But there is a need for more research to resolve the issues of the maintenance of effective photocatalysis and NP reusability associated with structural effects such as the porosity and surface area of metal oxides [70]. $NaYF_4$:(Gd, 1% Si)/TiO_2 is another well-known metal-oxide-coupled phosphor composite photocatalyst; it can be reused and is stable for up to three cycles, but the degradation efficiency of this material decreases from 95% to 60% and 40% in the 2nd and 3rd cycles, respectively. However, $NaYF_4$: (Gd, Si)/TiO_2 composites can be considered to be low-cost and efficient photocatalysts for the removal of pollutants from wastewater [71].

1.5 Factors affecting photocatalytic activity

1.5.1 The bandgap

In recent decades, wide-bandgap semiconductor photocatalysts have been assigned more importance due to their use in environmentally friendly wastewater treatment processes. Among the different types of oxides, ZnO and TiO_2 are the most promising materials; they have been studied many times because of their characteristics, such as high chemical stability, low cost, and nontoxic nature [72–74]. However, due to their large bandgap, these types of large-bandgap semiconductors can absorb only light in the UV region and are unable to absorb visible light. They also have a fast recombination rate of photogenerated electron–hole pairs that restricts the effective degradation of pollutants [75]. These issues related to semiconductor photocatalysts could be resolved if their light absorption range in the UV region could be easily extended to the visible region by tuning their bandgap; this might be made possible by adding some impurities to their structure or by adopting some new methods of synthesis [76]. In general, semiconductor photocatalysts bandgaps that fall between 1.23 and 3 eV include the oxidation and reduction potential ranges of H_2O. The redox potentials of the photocatalytic water splitting process correspond to $EH^+/H_2 = -4.44$ eV and $EO_2/H_2O = -5.67$ eV, respectively [77]; the difference between these is 1.23 eV, which is therefore the minimum bandgap that must be present for a material to be considered an effective photocatalyst. Fujishima and Honda reported water splitting by the photocatalytic approach; they studied both large- and narrow-bandgap semiconductor materials theoretically and experimentally [78, 79]. Semiconductor materials with a large bandgap (greater than 3 eV) have band edge positions suitable for the overall water splitting and hydrogen evolution processes, but their light absorption range is not compatible with the visible spectrum. Thus, there is a need for more efforts to search out new materials or design innovative materials with a broad absorption range that covers the full spectral component of visible light.

1.5.2 Particle size

The shape, size, and morphology of a material can affect its photocatalytic properties. Consider the bandgaps of bulk materials: the atomic orbitals overlap, producing bands with a small energy gap as compared to those of nanomaterials. However, in nanoscale metals, the orbitals are discontinuous; they rather form discrete energy levels in the band structure, which can be tuned by changing the nanoparticle diameter, as shown in figure 1.3. Hence, as the particle size decreases, the electrons become more confined in the particle, and confined electrons have more energy. Thus, the atomic orbitals of nanoparticles become discrete or quantized. The presence of the quantum confinement effect in nanomaterials leads to tunable electrical and optical properties.

Therefore, the bandgaps of bulk and nanostructured materials are different, and changes in particle size affect the bandgap energy, the light absorption capacity, and

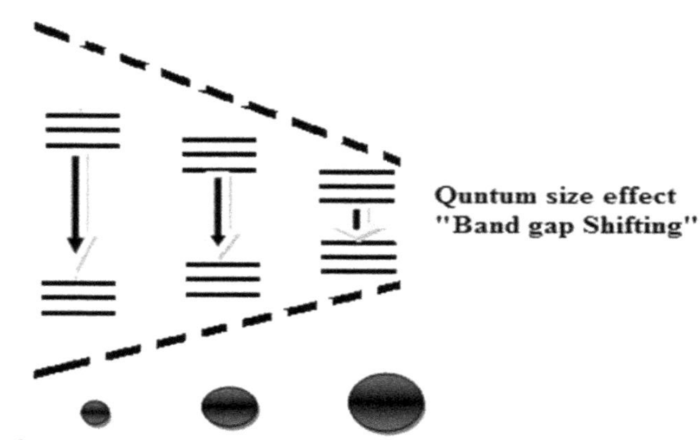

Figure 1.3. The effect of particle size on the bandgap.

the average free range of photogenerated charge carriers in the material. When the particle size of the material is smaller than the thickness of its space charge layer, which is negligible, then photogenerated charge carriers migrate from the bulk phase to the material's surface via the diffusion process and take part in surface redox reactions. The charge transfer rate of carriers and their separation efficiency can be improved by reducing the charge migration distance [80]. Therefore, small particles usually have a high specific area and show better adsorption properties that help to initiate the interaction between the catalyst and the reactants. Small particles thereby provide abundant active sites and a large light absorption area that hosts redox reactions and hence improves the photocatalytic performance [81].

1.5.3 Doping

Doping elements into a different host material is the most common strategy for improving the performance of metal-oxide semiconductors [82–85]. It can affect many parameters, such as the morphology, particle size, bandgap, binding energy, lattice defects, and other associated properties [86, 87]. For example, if the average particle size of pure SnO_2 particles is 52.3 nm, then the addition of Ag results in a reduction in the particle size of SnO_2; the new particle size may turn out to be 45.5 nm [88]. Further, Yakout *et al* [88] reported that the use of Co doping in an Ag/SnO_2 system was not effective in reducing the particle size. But the addition of higher amounts of dopants may affect the particle size, which may also result in a more uniform grain size distribution. Entradas *et al* also reported [89] that the particle size distribution was narrower when the amount of Co codoping was increased. This created agglomeration clusters of particles in pure SnO_2 due to the presence of the dopants [90].

1.6 Strategies for boosting photocatalytic efficiency

In the last few years, many researchers have made concerted efforts to design and develop new energy-efficient photocatalysts which are better at harvesting the maximum component of the solar spectrum, generating large numbers of charge carriers, and providing catalytic sites to support effective photocatalytic process [91]. To resolve the issues related to metal-oxide photocatalysts, various strategies have been proposed in the research literature to boost the photocatalytic efficiency of catalysts; some of these are discussed below.

1.6.1 Spatial separation of excitons

As discussed above, the photocatalytic efficiency of catalysts greatly depends on the separation rate of electron/hole pairs. In order to generate photoinduced charge carriers, a photocatalyst must be adequately excited by the incident photon flux energy. These positive and negative charge carriers can take femtoseconds (fs) to picoseconds (ps); they also subsequently take time to transit the surface of the photocatalyst (nanoseconds (ns) to microseconds (μs)) to reach the corresponding bands to initiate the redox reactions [92]. During the transit of charge carriers in the catalyst, there is a high probability of e^-/h^+ recombination that can release heat.

The recombination of e–/h+ pairs occurs within picoseconds to nanoseconds, which is much faster than their transfer rate to the surface of the catalyst (where they participate in oxidation and reduction reactions) [93]. Thus, to reduce the recombination rate, there is a need to develop new strategies to boost the separation of charge carriers and enhance the efficiency of photocatalysts.

1.6.1.1 The loading of cocatalysts

The loading of a cocatalyst improves the separation and transfer of excitons in the photocatalytic process, which helps to promote and stabilize the activity of the photocatalyst. Under light illumination, the photogenerated negative charge carriers in the CB of the catalyst are transferred toward the cocatalyst, which prevents the recombination of charge carriers in the VB [94]. Therefore, contact between the catalyst and the cocatalyst is essential for the transportation of charge carriers. When the charges reach the catalyst interface, the cocatalyst helps to improve the separation of charge carriers. When metallic cocatalysts are deposited on the surface of photocatalysts, they form a Schottky heterojunction that restricts the backward flow of electrons to the CB and thereby induces an electric field that enhances the separation of charge carriers, thereby promoting effective photocatalytic activity [95]. Semiconductor cocatalysts with narrow bandgaps also form heterostructures that improve charge separation. Similarly, cocatalysts consisting of transition-metal dichalcogenides have also exhibited better separation of electrons and holes because of their unique metallic and semiconductor structures. Among the different types of cocatalyst metals, cocatalysts such as Ag and Pt are excellent candidates for enhancing the photocatalytic efficiency of photocatalysts [96]. Recently, Sun *et al* [97] reported the photocatalytic response of CoP-loaded QDs as cocatalysts on CdS nanorods, confirming enhanced H_2 production under visible-light illumination.

1.6.1.2 Metal-oxide–coupled composites and their photocatalytic response

García *et al* [98] reported the photocatalytic performance of Bi codoped strontium aluminates blended with nanocrystalline TiO2. Here, small grains of TiO2 were made available on the surface of the strontium aluminate grains. With an increase in the surface area of the composite grains, an enhancement in photocatalytic activity was achieved. Figure 1.4 shows the photocatalytic degradation of MB achieved using the blended composite catalyst under UV light illumination. It can be seen that complete degradation of MB occurred after 210 min of UV exposure. Thus, the composite TiO_2–Bi codoped sample degraded 91.0% of the MB dye after 210 min of exposure time. Figure 1.4 shows that lower concentrations of Bi codoping in TiO_2–strontium aluminate composites exhibit a rapid photocatalytic degradation response as compared to pure TiO_2 powder.

1.6.1.3 Persistent phosphors and their photocatalytic response

Wang *et al* [99] reported the photocatalytic response of persistent phosphor used for the degradation of RhB. Figure 1.5 shows the persistent luminescence spectra of Zn^{2+} and Cr^{3+} codoped Ga_2O_3 phosphor. The photoluminescent excitation and emission spectra of the phosphor were observed in the UV–visible spectral range. When the

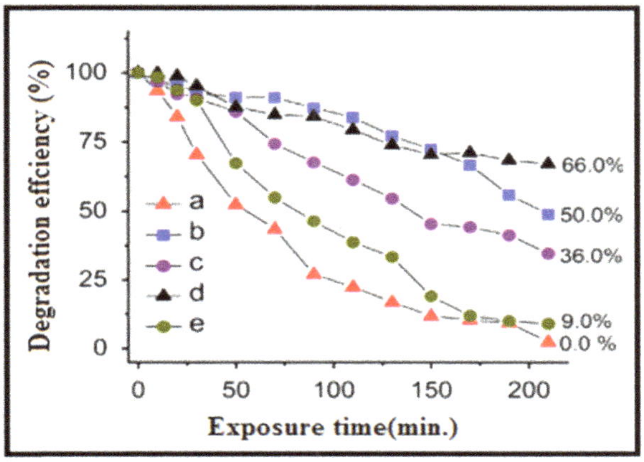

Figure 1.4. Photocatalytic degradation of MB as a function of exposure time under UV light illumination. The photodegradation responses of TiO_2–$Bi_{(2.0 \text{ and } 0.0 \text{ mol\%})}$ codoped strontium aluminate composites are shown in curves (a) and (b), and the response of pure TiO_2 is shown in curve (c); the responses of the TiO_2–$Bi_{(15.0 \text{ and } 1.0 \text{ mol\%})}$ codoped strontium aluminate composites are shown in curves (d) and (e), respectively. Reprinted from [98], Copyright (2015), with permission from Elsevier.

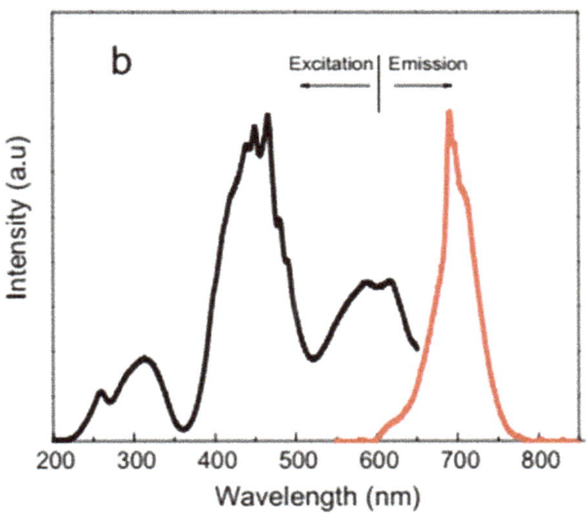

Figure 1.5. The persistent luminescence spectra of Ga_2O_3:$Cr^{3+}_{(0.01)}$ and $Zn^{2+}_{(0.005)}$ phosphors. Reprinted from [99], Copyright (2014), with permission from Elsevier.

phosphor is illuminated under ultraviolet light, a large number of charge carriers are created, and their energy is transferred to Cr^{3+} luminescence centers. These charge carriers are trapped by the oxygen vacancies, and after the irradiation source is cut off, the trapped electrons and holes are released and transferred to the luminescence centers by Cr^{3+} ions, which then emit characteristic persistent luminescence. Figure 1.6 shows the degradation of RhB as a function of the irradiation time.

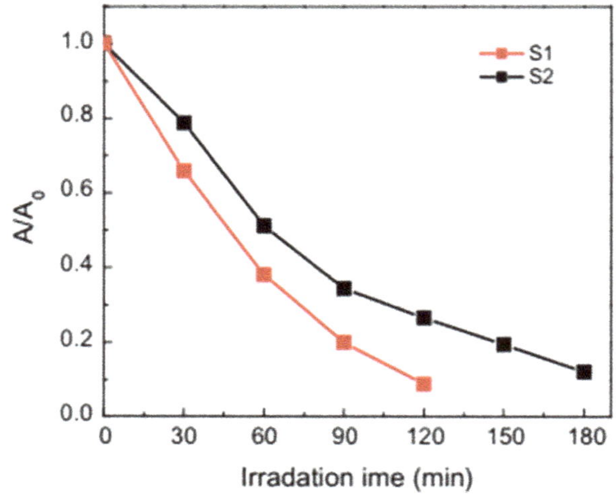

Figure 1.6. The absorption degradation of RhB as a function of irradiation time under ultraviolet light irradiation. Reprinted from [99], Copyright (2014), with permission from Elsevier.

It was found that the absorption of RhB photocatalyzed by $Ga_2O_3:Cr^{3+}_{(0.01)}$ reached 10% under UV light irradiation in 180 min, while the absorption of RhB photocatalyzed by the doping of $Zn^{2+}_{(0.005)}$ into $Ga_2O_3:Cr^{3+}_{(0.01)}$ took only 120 min. Thus, it can be seen that doping with Zn^{2+} can improve the photocatalytic properties of $Ga_2O_3:Cr^{3+}_{(0.01)}$ phosphor [100, 101].

1.6.1.4 The photocatalytic response of nanosized mixed metal oxides

It is well known that wide-bandgap nanosemiconductors such as TiO_2, ZnO, SnO_2, CeO_2, and NiO have excellent abilities to remove toxic dyes and organic pollutants from wastewater via photocatalytic degradation [102]. Nanosized titania (TiO_2) has a superior ability to remove textile dyes via degradation because of its effective generation of charge carriers under UV light illumination. However, due to its large bandgap, it does not work effectively under a visible-light source. Therefore, mixed metal-oxide systems have recently gained more importance due to their good photocatalytic degradation performance under visible or UV light sources. Various mixed metal-oxide systems that incorporate TiO_2, such as TiO_2–CeO_2, TiO_2–SnO_2, TiO_2–CuO, TiO_2–CdO, etc. have been shown to have excellent photocatalytic properties. These types of coupled metal-oxide systems exhibit better visible-light-driven photocatalytic activity and have higher dye degradation efficiency when used to remove toxic dyes and organic contaminants [103]. Recently, Rajendran *et al* [104] reported the photocatalytic response of a TiO_2/NiO composite catalyst used for the degradation of methyl orange. According to their experimental evidence, they observed 98% of methyl orange degradation within 60 min of irradiation. Here, the best performance (98%) of the composite catalyst was observed at pH = 7 (neutral), as shown in figure 1.7. In this composite system, the p–n junction takes a form in which Ni^{3+} states promote a large number of

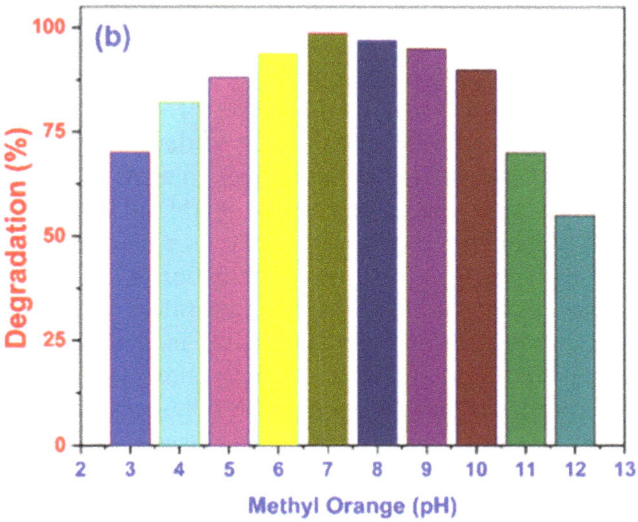

Figure 1.7. The degradation performance of the composite catalyst at different pH values. Reprinted from [104], Copyright (2020), with permission from Elsevier.

electrons, which reduces the recombination rate and helps to enhance the photocatalytic degradation process under visible light.

1.7 Applications

Due to globalization in the industrial sector, issues related to energy and the environment are becoming more serious. Nowadays, water and air pollution are hot topics due to their adverse effects on human health. To control and resolve these issues, there is a need to adopt sustainable technology for the betterment of mankind. Photocatalysis is an evergreen and economically viable way of solving the problems associated with wastewater treatment and air pollution. Hence, this research topic has become more attractive in the fields of science and engineering. Considering the global need for sustainable energy and environmental technology to replace traditional polluting technologies, photocatalysis is one of the better approaches with which to explore the innovative idea of using clean and natural solar light energy [105]. In the 1970s, Honda and Fujishima published their important discoveries related to water splitting and hydrogen generation using TiO_2 semiconductors for photocatalysis [60]. The scope of photocatalysts in energy and environmental sustainability is discussed below.

1.7.1 Energy sustainability

In the last few decades, many countries have used fossil fuels such as coal, oil, and agricultural waste products to generate electricity in power plants to fulfill the demand for energy in all sectors; such fossil fuel uses have made a major contribution to air, water, and soil pollution all over the world. The utilization of oil and petroleum products in automobiles and heavy transportation vehicles is another major cause of air pollution. The use of this type of traditional power generation technology disperses many kinds of harmful pollutants into the atmosphere, so the level of toxic contaminants affecting the air quality index is increasing rapidly; these pollutants are also a major cause of global warming. To reduce environmental damage, there is an urgent need to develop and replace the traditional sources of energy and related polluting technologies with sustainable energy sources. Among the different renewable and sustainable energy sources, hydrogen is the most promising source for the current century, as it has the advantages of being an eco-friendly form of energy with lower production costs; however, it still has some challenges, such as increasing the production rate, storing the fuel, etc [106]. The photocatalytic production of hydrogen results in pure hydrogen that can be converted into energy and H_2O, which is also environmentally friendly [107]. In the past, hydrogen was produced using nonrenewable resources such as natural gas and petroleum-based technologies. However, these processes suffer from some disadvantages and liberate some other pollutants that are not much cheaper to deal with from a commercial perspective [108]. Hence, considering the previous drawbacks as well as the economic and environmental benefits of the new and advanced photocatalytic hydrogen generation process using solar light energy as the source of clean and low-cost sustainable energy technologies [109].

1.7.2 Environmental sustainability

In recent years, water pollution from industrial waste has been a very serious global issue. Contamination due to heavy metals and organic dye molecules in natural water resources accumulates for a long time, causing negative effects for organisms and human health.

To conserve natural water resources and control environmental pollution, there is an urgent need to develop environmentally friendly water purification technology and other sustainable energy technologies to avoid environmental issues in the air, water, and soil. From the literature database [100–110], it can be seen that approximately 10 000 types of dyes are used in industry for different purposes; therefore, the amounts of textile dyes and other dyes found in industrial wastewater are excessive. When these spread into fresh water resources, many types of toxic contamination can affect the quality of the water. The consumption and use of such contaminated water may cause health issues such as skin rashes, sinus infections, and cancer by entering the human ecosystem through water and animals [110]. Traditional methods including biological treatment, reverse osmosis, coagulation, adsorption, and ultrafiltration are ineffective water treatment processes [111]. Compared to these methods, the energy-efficient photocatalysis process has great potential to remove and degrade organic pollutants naturally using solar energy; it represents the lowest-cost and most favorable method that can utilize low-cost and nontoxic catalysts for the development of technology for water purification and environmental protection [112, 113]. Therefore, many research organizations are currently working in fundamental and applied research areas in the field of environmental sustainability to protect our planet and maintain the harmony of nature.

1.8 Conclusions and future prospects

Photocatalysis is a promising low-cost, energy-efficient, and environmentally friendly technique for resolving environmental and energy problems. There is a great demand for metal-oxide photocatalysts with a visible-light-driven photocatalytic response for the production of hydrogen and wastewater treatment. Further, research into large-scale production and treatment processes remains incomplete. So, in the future, more work is needed to develop this environmentally friendly technology that utilizes natural sunlight for the photocatalytic reaction. In this century, more research is focused on the development of sustainable technology, and photocatalysis has attracted great interest due to its attractive and broad scope for use in the near future due to the current global problems related to energy and the environment. There are still a few challenges in the development of photocatalytic materials, such as achieving high efficiency, thermal stability, purity, and environmental friendliness and overcoming low efficiency in photocatalytic reactors, which is also a major hurdle for their use on a large scale.

References

[1] Briffa J, Sinagra E and Blundell R 2020 Heavy metal pollution in the environment and their toxicological effects on humans *Heliyon* **6** E04691

[2] Hisatomi T and Domen K 2019 Reaction systems for solar hydrogen production via water splitting with particulate semiconductor photocatalysts *Nat. Catal.* **2** 387–99

[3] Kosco J *et al* 2020 Enhanced photocatalytic hydrogen evolution from organic semiconductor heterojunction nanoparticles *Nat. Mater.* **19** 559–65

[4] Meng X, Wang S, Zhang C, Dong C, Li R, Li B, Wang Q and Ding Y 2022 Boosting Hydrogen Evolution Performance of a CdS-Based Photocatalyst: In Situ Transition from Type I to Type II Heterojunction during Photocatalysis *ACS Catal.* **12** 10115–26

[5] Zhao X, Li J, Kong X, Li C, Lin B, Dong F, Yang G, Shao G and Xue C 2022 Carbon Dots Mediated In Situ Confined Growth of Bi Clusters on g-C_3N_4 Nanomeshes for Boosting Plasma-Assisted Photoreduction of CO_2 *Small* **18** 2204154

[6] Zhou Q, Guo Y, Ye Z, Fu Y, Guo Y and Zhu Y 2022 Carbon nitride photocatalyst with internal electric field induced photogenerated carriers spatial enrichment for enhanced photocatalytic water splitting *Mater. Today* **58** 100–9

[7] Collado L, Naranjo T, Gomez-Mendoza M, López-Calixto C, Oropeza F, Liras M, Marugán J and Peña O'Shea V 2021 Conjugated Porous Polymers Based on BODIPY and BOPHY Dyes in Hybrid Heterojunctions for Artificial Photosynthesis *Adv. Funct. Mater.* **31** 2105384

[8] Chen J, Tang T, Feng W, Liu X, Yin Z, Zhang X, Chen J and Cao S 2022 Largescale synthesis of p–n heterojunction Bi_2O_3/TiO_2 nanostructures as photocatalysts for removal of antibiotics under visible light *ACS Appl. Nano Mater.* **5** 1296–307

[9] Khan M M 2023 Chapter 8—Photocatalysis: laboratory to market *Theoretical Concepts of Photocatalysis* ed M Mansoob Khan (Amsterdam: Elsevier) 187–212

[10] Tong H, Ouyang S, Bi Y, Umezawa N, Oshikiri M and Ye J 2012 Nano-photocatalytic materials: possibilities and challenges *Adv. Mater.* **24** 229–51

[11] Ihsan A, Irshad A, Warsi M F, Din M I and Zulfiqar S 2022 $NiFe_2O_4$/ZnO nanoparticles and its composite with flat 2D rGO sheets for efficient degradation of colored and colorless effluents photocatalytically *Opt. Mater.* **134** 113213
[12] Irshad A, Farooq F, Warsi M F, Shaheen N, Elnaggar A Y, Hussein E E, ElBahy Z M and Shahid M 2022 Ag-doped $FeCo_2O_4$ nanoparticles and their composite with flat 2D reduced graphene oxide sheets for photocatalytic degradation of colored and colorless compounds *Flat Chem.* **31** 100325
[13] Irshad A, Warsi M F, Agboola P O, Dastgeer G and Shahid M 2022 Sol–gel assisted Ag doped $NiAl_2O_4$ nanomaterials and their nanocomposites with g-$C3N_4$ nanosheets for the removal of organic effluents *J. Alloys Compd.* **902** 163805
[14] Santhi K, Manikandan P, Rani C and Karuppuchamy S 2015 Synthesis of nanocrystalline titanium dioxide for photodegradation treatment of remazol Brown dye *Appl. Nanosci.* **5** 373–8
[15] Matthews R W 1988 An adsorption water purifier with *in situ* photocatalytic regeneration *J. Catal.* **113** 549–55
[16] Henderson M A, Epling W S, Perkins C L, Peden C H F and Diebold U J 1999 Interaction of molecular oxygen with the vacuum-annealed TiO_2 (110) surface: molecular and dissociative channels *Phys. Chem.* B **103** 5328–37
[17] Karthikeyan C, Thamima M and Karuppuchamy S 2019 Structural and photocatalytic property of $CaTiO_3$ nanosphere *Mater. Sci. Forum* **979** 169–74
[18] Vinodgopal K, Stafford U, Gray K A and Kamat P V 1994 The role of oxygen and reaction intermediates in the degradation of 4-chlorophenol on immobilized TiO_2 particulate films *J. Phys. Chem.* **98** 6797–803
[19] Zhang X Y, Ling S Y, Ji H Y, Xu L, Huang Y, Hua M Q, Xia J X and Li H M 2017 Metal ion-containing ionic liquid assisted synthesis and enhanced photoelectrochemical performance of g-$C3N_4$/ZnO composites *Mater. Technol.* **3** 1–8
[20] Lin N, Gong Y, Wang R, Wang Y and Zhang X 2022 Critical review of perovskite-based materials in advanced oxidation system for wastewater treatment: design, applications and mechanisms *J. Hazard. Mater.* **424** 127637
[21] Verma V, Al-Dossari M, Singh J, Rawat M, Kordy M G M and Shaban M 2022 A review on green synthesis of TiO_2 NPs: photocatalysis and antimicrobial applications *Polym* **14** 1444
[22] Mittal M, Dana J, Lübkemann F, Ghosh H N, Bigall N C and Sapra S 2022 Insight into morphology dependent charge carrier dynamics in ZnSe–CdS nanoheterostructures *Phys. Chem. Chem. Phys.* **24** 8519–28
[23] Alfryyan N, Kordy M G M, Abdel-Gabbar M, Soliman H A and Shaban M 2022 Characterization of the biosynthesized intracellular and extracellular plasmonic silver nanoparticles using *Bacillus cereus* and their catalytic reduction of methylene blue *Sci. Rep.* **121** 14
[24] Pawar R C and Lee C S 2015 Basics of photocatalysis *Heterogeneous Nanocomposite-Photocatalysis for Water Purification* ed R C Pawar and C S Lee (Boston, MA: William Andrew Publishing) 1–23 ch 1
[25] Feng X, Hu G and Hu J 2011 Solution-phase synthesis of metal and/or semiconductor homojunction/heterojunction nanomaterials *Nanoscale* **3** 2099–117

[26] Li R, Weng Y, Zhou X, Wang X, Mi Y, Chong R, Han H and Li C 2015 Achieving overall water splitting using titanium dioxide-based photocatalysts of different phases *Energy Environ. Sci.* **8** 2377–82
[27] Zheng C, Huang L, Guo Q, Chen W, Li W and Wang H 2018 Facile one-step fabrication of upconversion fluorescence carbon quantum dots anchored on graphene with enhanced nonlinear optical responses *RSC Adv.* **8** 10267–76
[28] Martha S, Mansingh S, Parida K M and Thirumurugan A 2017 Exfoliated metal free homojunction photocatalyst prepared by a biomediated route for enhanced hydrogen evolution and Rhodamine B degradation *Mater. Chem. Front.* **1** 1641–53
[29] Navarrete-Magana M, Estrella-Gonzalez A, May-Ix L, Cipagauta-Diaz S and Gomeza R 2021 Improved photocatalytic oxidation of arsenic (III) with WO_3/TiO_2 nanomaterials synthesized by the sol-gel method *J. Environ. Manage.* **282** 111602
[30] Majhi D, Das K, Bariki R, Padhan S, Mishra A, Dhiman R, Dash P, Nayakc B and Mishra B G 2020 A facile reflux method for in situ fabrication of a non-cytotoxic $Bi_2S_3/\beta\text{-}Bi_2O_3/ZnIn_2S_4$ ternary photocatalyst: a novel dual Z-scheme system with enhanced multifunctional photocatalytic activity *J. Mater. Chem.* A **8** 21729–43
[31] Cheng T T, Gao H J, Sun X F, Xian T, Wang S F, Yi Z, Liu G R, Wang X X and Yang H 2021 An excellent Z-scheme $Ag_2MoO_4/Bi_4Ti_3O_{12}$ heterojunction photocatalyst: Construction strategy and application in environmental purification *Adv. Powder Technol.* **32** 951–62
[32] Chen F, Ma T, Zhang T, Zhang Y and Huang H 2021 Atomic-level charge separation strategies in semiconductor-based photocatalysts *Adv. Mater.* **33** 2005256
[33] Qiao B, Wang A, Yang X, Allard L F, Jiang Z, Cui Y, Liu J, Li J and Zhang T 2011 Single-atom catalysis of CO oxidation using Pt1/FeOx *Nat. Chem.* **3** 634–41
[34] Li X, Yang X, Huang Y, Zhang T and Liu B 2019 Supported noble-metal single atoms for heterogeneous catalysis *Adv. Mater.* **31** 1902031
[35] Shan J, Li M, Allard L F, Lee S and Flytzani-Stephanopoulos M 2017 Mild oxidation of methane to methanol or acetic acid on supported isolated rhodium catalysts *Nature* **551** 605–8
[36] Qin R, Liu K, Wu Q and Zheng N 2020 Surface coordination chemistry of atomically dispersed metal catalysts *Chem. Rev.* **120** 11810–99
[37] Ji S, Chen Y, Wang X, Zhang Z, Wang D and Li Y 2020 Chemical synthesis of single atomic site catalysts *Chem. Rev.* **120** 11900–55
[38] Wang A, Li J and Zhang T 2018 Heterogeneous single-atom catalysis *Nat. Rev. Chem.* **2** 65–81
[39] Jiao L, Yan H, Wu Y, Gu W, Zhu C, Du D and Lin Y 2020 When nanozymes meet single-atom catalysis *Angew. Chem.* **132** 2585–96
[40] Gao C, Low J, Long R, Kong T, Zhu J and Xiong Y 2020 Heterogeneous single-atom photocatalysts: fundamentals and applications *Chem. Rev.* **120** 12175–216
[41] Yi D, Lu F, Zhang F, Liu S, Zhou B, Gao D, Wang X and Yao J 2020 Regulating charge transfer of lattice oxygen in single-atom-doped titania for hydrogen evolution *Angew. Chem. Int. Ed.* **59** 15855–9
[42] Feng H *et al* 2022 Porphyrin-based Ti-MOFs conferred with single-atom Pt for enhanced photocatalytic hydrogen evolution and NO removal *Chem. Eng. J.* **428** 132045
[43] Apte S K, Garaje S N, Naik S D, Waichal R P and Kale B B 2013 Environmentally benign enhanced H_2 production from abundant copious waste H_2S using size tuneable cubic bismuth (Bi0) quantum dots-GeO_2 glass photocatalyst under solar light *Green Chem.* **15** 3459–67

[44] Zhao C, Li W, Liang Y, Tian Y and Zhang Q 2016 Synthesis of BiOBr/carbon quantum dots microspheres with enhanced photoactivity and photostability under visible light irradiation *Appl. Catal. Gen.* **527** 127–36
[45] Murphy C J 2002 Peer reviewed: optical sensing with quantum dots *Anal. Chem.* **74** 520A–6A
[46] Kandi D, Martha S and Parida K M 2017 Quantum dots as enhancer in photocatalytic hydrogen evolution: a review *Int. J. Hydrogen Energy* **42** 9467–81
[47] Das N and Kandimalla S 2017 Application of perovskites towards remediation of environmental pollutants: an overview *Int. J. Environ. Sci. Technol.* **147** 1559–72
[48] Mahmoudi F, Saravanakumar K, Maheskumar V, Njaramba L K, Yoon Y and Park C M 2022 Application of perovskite oxides and their composites for degrading organic pollutants from wastewater using advanced oxidation processes: review of the recent progress *J. Hazard. Mater.* **436** 129074
[49] Xu Q L, Zhang L Y, Cheng B, Fan J J and Yu J G 2020 S-Scheme Heterojunction Photocatalyst *Chem.* **6** 1543–59
[50] Liu Y, Hao X Q, Hu H Q and Jin Z L 2021 High Efficiency Electron Transfer Realized over NiS_2/$MoSe_2$ S-Scheme Heterojunction in Photocatalytic Hydrogen Evolution *Acta Phys.-Chim. Sin.* **37** 2008030
[51] Jia X M, Hu C, Sun H Y, Cao J, Lin H L, Li X Y and Chen S F 2023 A dual defect co-modified S-scheme heterojunction for boosting photocatalytic CO_2 reduction coupled with tetracycline oxidation *Appl. Catal.,* B **324** 122232
[52] Zhang X, Chen Z, Li X M, Wu Y, Zheng J F, Li Y Q, Wang D B, Yang Q, Duan A B and Fan Y C 2023 Promoted electron transfer in Fe^{2+}/Fe^{3+} co-doped $BiVO_4$/Ag_3PO_4 S-scheme heterojunction for efficient photo-Fenton oxidation of antibiotics *Sep. Purif. Technol.* **310** 123116
[53] Li Y F, Xia Z L, Yang Q, Wang L X and Xing Y 2022 Review on g-C_3N_4-based S-scheme heterojunction photocatalysts *J. Mater. Sci. Technol.* **125** 128–44
[54] Zhang X D, Yu J G, Macyk W, Wageh S, Al-Ghamdi A A and Wang L X 2023 C_3N_4/PDA S-Scheme Heterojunction with Enhanced Photocatalytic H_2O_2 Production Performance and Its Mechanism *Adv. Sustain. Syst.* **7** 2200113
[55] Chu H, Lee C and Tai N 2016 Green preparation using black soybeans extract for graphene-based porous electrodes and their applications in supercapacitors *J. Power Sources* **322** 31–9
[56] Tarcan R, Todor-Boer O, Petrovai I, Leordean C, Astilean S and Botiz I 2020 Reduced graphene oxide today *J. Mater. Chem.* C **8** 1198–224
[57] Witjaksono G *et al* 2021 Effect of nitrogen doping on the optical bandgap and electrical conductivity of nitrogen-doped reduced graphene oxide *Molecules* **26** 6424
[58] Ngidi N P D, Ollengo M A and Nyamori V O 2020 Tuning the properties of boron-doped reduced graphene oxide by altering the boron content *New J. Chem.* **44** 16864–76
[59] Hu C, Lu T, Chen F and Zhang R 2013 A brief review of graphene-metal oxide composites synthesis and applications in photocatalysis *J. Chin. Adv. Mater. Soc.* **1** 21–39
[60] Fujishima A and Honda K 1972 Electrochemical photolysis of water at a semiconductor electrode *Nature* **238** 37–8
[61] Yoshida M, Yamakata A, Takanabe K, Kubota J, Osawa M and Domen K 2009 ATR–SEIRAS investigation of the Fermi level of Pt cocatalyst on a GaN photocatalyst for hydrogen evolution under irradiation *J. Am. Chem. Soc.* **131** 13218
[62] Mittal M, Sharma M and Pandey O P 2016 Fast and quick degradation properties of doped and capped ZnO nanoparticles under UV–visible light radiations *Sol. Energy* **125** 51–64

[63] Cho S, Jang J W, Kim J, Lee J S, Choi W and Lee K H 2011 Three-dimensional type II ZnO/ZnSe heterostructures and their visible light photocatalytic activities *Langmuir* **27** 10243–50

[64] Varghese Alex K, Tamil Pavai P, Rugmini R, Shiva Prasad M, Kamakshi K and Sekhar K C 2020 Green synthesized Ag nanoparticles for bio-sensing and photocatalytic applications *ACS Omega* **5** 13123–9

[65] Akter S, Lee S Y, Siddiqi M Z, Balusamy S R, Ashrafudoulla M, Rupa E J and Huq M A 2020 Ecofriendly synthesis of silver nanoparticles by *Terrabacter humi* sp. nov. and their antibacterial application against antibiotic-resistant pathogens *Int. J. Mol. Sci.* **21** 9746

[66] Garba Z N, Zhou W, Zhang M and Yuan Z 2020 A review on the preparation, characterization and potential application of perovskites as adsorbents for wastewater treatment *Chemosphere* **244** 125474

[67] Kafle B 2020 Introduction to nanomaterials and application of UV–visible spectroscopy for their characterization *Chemical Analysis and Material Characterization by Spectrophotometry* (Amsterdam: Elsevier) 6 147–98

[68] Huston M, DeBella M, DiBella M and Gupta A 2021 Green synthesis of nanomaterials *Nanomaterials* **11** 2130

[69] Shivashankar A, Prashantha S, Anantharaju K, Malini S, Manjunatha H, Vidya Y, Sridhar K and Munirathnam R 2022 Rod shaped zirconium titanatenanoparticles: synthesis, comparison and systematic investigation of structural, photoluminescence, electrochemical sensing and supercapacitor properties *Ceram. Int.* **48** 35676

[70] Shan A Y, Ghazi T I M and Rashid S A 2010 Immobilisation of titanium dioxide onto supporting materials in heterogeneous photocatalysis: a review *Appl. Catal. A Gen.* **389** 1–8

[71] Mavengere S and Kim J-S 2018 UV–visible light photocatalytic properties of $NaYF_4$:(Gd, Si)/TiO_2 composites *Appl. Surf. Sci.* **444** 491–6

[72] Zheng Y, Zheng L, Zhan Y, Lin X, Zheng Q and Wei K 2007 Ag/ZnO heterostructure nanocrystals: synthesis, characterization, and photocatalysis *Inorg. Chem.* **46** 6980–6

[73] Wu H B, Hng H H and Lou X W D 2012 Direct synthesis of anatase TiO_2 nanowires with enhanced photocatalytic activity *Adv. Mater.* **24** 2567–71

[74] Liu Z, Zhang X, Nishimoto S, Jin M, Tryk D A, Murakami T and Fujishima A 2007 Highly ordered TiO2 nanotube arrays with controllable length for photoelectrocatalytic degradation of phenol *J. Phys. Chem.* C **112** 253–9

[75] Fan T, Han T, Chow S K and Zhang D 2010 Biogenic N–P-codoped TiO_2: synthesis, characterization and photocatalytic properties *Bioresour. Technol.* **101** 6829–35

[76] Bora T, Sathe P, Laxman K, Dobrestov S and Dutta J 2017 Defect engineered visible light active ZnO nanorods for photocatalytic treatment of water *Catal. Today* **284** 11–8

[77] Jiao Y, Zhou L, Ma F, Gao G, Kou L, Bell J, Sanvito S and Du A 2016 Predicting Single-Layer Technetium Dichalcogenides (TcX_2, X = S, Se) with Promising Applications in Photovoltaics and Photocatalysis *ACS Appl. Mater. Interfaces* **8** 5385–92

[78] Fujishima A and Honda K 1972 Electrochemical Photolysis of Water at a Semiconductor Electrode *Nature* **238** 37–8

[79] Zhuang H L and Hennig R G 2013 Single-Layer Group-III Monochalcogenide Photocatalysts for Water Splitting *Chem. Mater.* **25** 3232–8

[80] Cao Y, Gao Q, Li Q, Jing X, Wang S and Wang W 2017 Synthesis of 3D porous MoS_2/g-C_3N_4 heterojunction as a high efficiency photocatalyst for boosting H_2 evolution activity *RSC Adv.* **7** 40727–33

[81] Gordon T R, Cargnello M, Paik T, Mangolini F, Weber R T, Fornasiero P and Murray C B 2012 Nonaqueous synthesis of TiO_2 nanocrystals using TiF_4 to engineer morphology, oxygen vacancy concentration, and photocatalytic activity *J. Am. Chem. Soc.* **134** 6751–61

[82] Yang Y, Wang S-Q, Wen H, Ye T, Chen J, Li C-P and Du M 2019 Nanoporous gold embedded ZIF composite for enhanced electrochemical nitrogen fixation *Angew. Chem. Int. Ed.* **58** 15362–6

[83] Zhang L, Huang J, Hu Z, Li X, Ding T, Hou X, Chen Z, Ye Z and Luo R 2022 Ni $(NO_3)_2$-induced high electrocatalytic hydrogen evolution performance of self-supported fold-like WC coating on carbon fiber paper prepared through molten salt method *Electrochim, Acta* **422** 140553

[84] Zhao C, Xi M, Huo J and He C 2021 B-doped 2D-InSe as a bifunctional catalyst for CO_2/CH_4 separation under the regulation of an external electric field *Phys. Chem. Chem. Phys.* **23** 23219–24

[85] Zhang L, Hu Z, Huang J, Chen Z, Li X, Feng Z, Yang H, Huang S and Luo R 2022 Experimental and DFT studies of flower-like Ni-doped Mo_2C on carbon fiber paper: a highly efficient and robust HER electrocatalyst modulated by $Ni(NO_3)_2$ concentration *J. Adv. Ceram.* **11** 1294–306

[86] Wang D, Wang X-X, Jin M L, He P and Zhang S 2022 Molecular level manipulation of charge density for solid–liquid TENG system by proton irradiation *Nano Energy* **103** 107819

[87] Yang B, Liu T, Guo H, Xiao S and Zhou L 2019 High-performance meta-devices based on multilayer meta-atoms: interplay between the number of layers and phase coverage *Sci. Bull.* **64** 823–35

[88] Yakout S M 2019 Inclusion of cobalt reinforced Ag doped SnO_2 properties: electrical, dielectric constant, magnetic and photocatalytic insights *J. Mater. Sci., Mater. Electron.* **30** 17053–65

[89] Entradas T, Cabrita J F, Dalui S, Nunes M R, Monteiro O C and Silvestre A J 2014 Synthesis of sub-5 nm Co-doped SnO_2 nanoparticles and their structural, microstructural, optical and photocatalytic properties *Mater. Chem. Phys.* **147** 563–71

[90] Nihal S, Rattan M, Anjali H, Kumar S, Sharma M, Tripathi S K and Goswamy J K 2021 Synthesis and characterization of Ag metal doped SnO_2, WO_3 and WO_3–SnO_2 for propan-2-ol sensing *Results Mater.* **9** 100127

[91] Kuspanov Z, Bakbolat B, Baimenov A, Issadykov A, Yeleuov M and Daulbayev C 2023 Photocatalysts for a sustainable future: innovations in large-scale environmental and energy applications *Sci. Total Environ.* **885** 163914–24

[92] Wang J, Wang Z, Dai K and Zhang J 2023 Review on inorganic–organic S-scheme photocatalysts *J. Mater. Sci. Technol.* **165** 187–218

[93] Sahai A, Ikram S, Rai R, Shrivastav S, Dass V R and Satsangi 2017 Quantum dots sensitization for photoelectrochemical generation of hydrogen: a review *Renew. Sustain. Energy Rev.* **68** 19–27

[94] Xiong Z, Lei Z, Ma S, Chen X, Gong B and Zhao Y 2017 Photocatalytic CO_2 reduction over V and W codoped TiO_2 catalyst in an internal-illuminated honeycomb photoreactor under simulated sunlight irradiation *Appl Catal Environ* **219** 412–24

[95] Liu X, Chen X, Wang S, Yan L, Yan J and Guo H 2022 Promoting the photocatalytic H_2 evolution activity of $CdLa_2S_4$ nanocrystalline using few-layered WS_2 nanosheet as a co-catalyst *Int. J. Hydrogen Energy* **47** 2327–37

[96] Zhu Y, Wang T, Xu T, Li Y and Wang C 2019 Size effect of Pt co-catalyst on photocatalytic efficiency of g-C_3N_4 for hydrogen evolution *Appl. Surf. Sci.* **464** 36–42
[97] Sun Q, Yu Z, Jiang R, Hou Y, Sun L and Qian L 2020 CoP QD anchored carbon skeleton modified CdS nanorods as a co-catalyst for photocatalytic hydrogen production *Nanoscale* **12** 19203–12
[98] García C R, Diaz-Torres L A, Salas P, Guzman M and Angeles-Chavez C 2015 Photoluminescent and photocatalytic properties of bismuth doped strontium aluminates blended with titanium dioxide *Mater. Sci. Semicond. Process.* **37** 105–11
[99] Wang Y, Xu K, Li D, Zhao H and Hu Z 2014 Persistent luminescence and photocatalytic properties of Ga_2O_3:Cr^{3+}, Zn^{2+} phosphors *Opt. Mater.* **36** 1798–801
[100] Sato J, Kobayashi H, Zkarashi K, Saito N, Nishiyama H and Inoue Y 2004 *J. Phys. Chem.* B **108** 4369–75
[101] Girija K, Thirumalairajan S, Avadhani G S, Mangalaraj D, Ponpandian N and Viswanathan C 2013 *Mater. Res. Bull.* **48** 2296–303
[102] Serpone N and Emeline A V 2012 Semiconductor photocatalysis-past, present, and future outlook *J. Phys. Chem. Lett.* **35** 673–7
[103] Lu D, Zelekew O A, Abay A K, Huang Q, Chen X and Zheng Y 2019 Synthesis and photocatalytic activities of a CuO/TiO_2 composite catalyst using aquatic plants with accumulated copper as a template *RSC Adv.* **9** 2018–25
[104] Rajendran S, Manoj D, Nimita Jebaranjitham J, Kumar B G, Bharath G, Banat F, Qin J, Vadivel S and Gracia F 2020 Nanosized titania-nickel mixed oxide for visible light photocatalytic activity *J. Mol. Liq.* **311** 113328
[105] Kudo A and Miseki Y 2009 Heterogeneous photocatalyst materials for water splitting *Chem. Soc. Rev.* **38** 253–78
[106] Dawood F, Anda M and Shafiullah G M 2020 Hydrogen production for energy: an overview *Int. J. Hydrogen Energy* **45** 3847–69
[107] Holladay J D, Hu J, King D L and Wang Y 2009 An overview of hydrogen production technologies *Catal. Today* **139** 244–60
[108] Li Y and Tsang S C E 2020 Recent progress and strategies for enhancing photocatalytic water splitting *Mater. Today Sustain.* **9** 100032
[109] Armaroli N and Balzani V 2010 The Hydrogen Issue *ChemSusChem.* **4** 21–36
[110] Hariharan D 2020 Enhanced photocatalysis and anticancer activity of green hydrothermal synthesized Ag@ TiO_2 nanoparticles *J. Photochem. Photobiol.* B **202** 111636
[111] Ahmed M A *et al* 2020 Rapid photocatalytic degradation of RhB dye and photocatalytic hydrogen production on novel curcumin/SnO_2 nanocomposites through direct Z-scheme mechanism *J. Mater. Sci., Mater. Electron.* **31** 19188
[112] Hamdy M S *et al* 2021 Fabrication of novel polyaniline/ZnO heterojunction for exceptional photocatalytic hydrogen production and degradation of fluorescein dye through direct Z-scheme mechanism *Opt. Mater.* **117** 111198
[113] Sayed M A *et al* 2022 Mesoporous polyaniline/SnO_2 nanospheres for enhanced photocatalytic degradation of bio-staining fluorescent dye from an aqueous environment *Inorg. Chem. Commun.* **139** 109326

IOP Publishing

Photocatalysts for Energy and Environmental Sustainability

Vijay B Pawade and Bharat A Bhanvase

Chapter 2

Rare-earth-doped metal oxide photocatalysts for wastewater treatment

G A Suganya Josephine, S Rubesh Ashok Kumar and D Vasvini Mary

Semiconductor metal oxide photocatalysts are well-known candidates for wastewater treatment. The enhancement of photocatalytic activity is often attempted using the approach of doping metals and nonmetals. In recent years, rare-earth (RE) materials have explored. These materials are characterized by the presence of f *orbitals*, which play a vital role in bandgap modification in doped materials. The *f–f transitions* contribute to electron trapping, thereby enhancing the lifetime of free holes and accelerating the photocatalytic process. This chapter explores the various rare-earth dopants and the doping process employed for nanomaterial preparation. The most important application of these nanomaterials is for the treatment of wastewater under UV, visible light, and natural solar irradiation, which are further discussed.

2.1 Introduction

Global environmental degradation results from intensified industry brought on by population growth and development. Industries pollute the environment by discharging heavy metal ions and harmful chemicals into waterways [1]. Therefore, protecting the environment from the damaging effects of discarded chemicals and heavy metal ions is crucial. Wastewater containing organic pollutants cannot be treated efficiently using traditional supplies or methods [2, 3]. In order to provide safe and clean water, the water industry needs to use low-cost, sustainable, efficient, and effective water treatment processes [4]. Numerous industrial applications and water treatment methods successfully utilize nanotechnology [5]. As one of the many nanotechnology-based remediation methods, semiconductor-mediated heterogeneous advanced oxidation technologies have become a viable option. As part of advanced oxidation processes (AOPs), extremely reactive hydroxyl radicals ($^{\bullet}OH$)

doi:10.1088/978-0-7503-5697-8ch2

are produced on-site and used to turn various organic contaminants into non-hazardous molecules [6–8].

Semiconductor metal oxides are a class of materials widely used for various applications in the field of science and technology. Almost all areas of research and technology have used semiconductor nanoparticles. Researchers from all over the world have been studying semiconductors at the nanoscale, and their use in heterogeneous photocatalysis has become increasingly important in biomedical nanotechnology. Although TiO_2 nanoparticles are superior semiconductors for photocatalytic applications, they are expensive [9]. In order to replace TiO_2, a wide range of metal oxides have been researched, including ZnO, SnO_2, Bi_2O_3, WO_3 and other metal oxides [10]. When exposed to UV/visible light, surface changes occur that prevent the recombination of electron–hole pairs, which is necessary for the generation of radical species [11].

Researchers from all over the world have tried to modify pure metal oxides by doping them with metals, nonmetals, and RE materials; by producing composites and core–shell materials; by supporting metal oxides on SiO_2, Al_2O_3, graphene oxide, or other materials; etc [12]. In the periodic table, the lanthanides are found among the RE elements. The lanthanides consist of 15 different elements, ranging from lutetium (71) to lanthanum (57), and the f *orbitals* in these compounds are mainly responsible for these materials' optical and catalytic capabilities [13]. In the years after Carl Axel Arrhenius discovered a black mineral, 'ytterbite,' in 1787, the REs gradually become known to the world. Lanthanides are mostly used as up-conversion catalysts. They are also added to metal oxides as dopants to improve their characteristics [14]. In addition to their function as doping agents, these substances are also employed to create composites and core–shell materials.

The luminescence of the metal oxide ZnO is improved when lanthanide (RE^{3+}) ions are used to modify its surface. The metal oxide with embedded lanthanide converts near-IR and visible radiation into UV wavelengths, which results in the sequential absorption of numerous infrared photons. The lanthanides' partially filled f-*orbitals* or f-subshells contribute to the *f–f transition*, which is what causes the luminescent property. Since the RE ion has higher photocatalytic activity, it appears that these transitions can be viewed as narrow emission bands that are unique to it [15]. This chapter covers: (i) types of RE material that have been employed to enhance the photocatalytic activity of metal oxides and (ii) REs employed as photocatalysts under UV light or visible light. The preparation techniques employed, the morphology of the materials, the pollutants of interest, and the percentage of degradation are detailed in the following sections.

2.2 Semiconductor metal oxides

The photocatalytic splitting of water by TiO_2 under visible light was first performed in 1972 by Fujishima and Honda. Since then, research into the preparation of an alternative semiconductor metal oxide for use in photocatalysis has been strongly encouraged [16, 17]. TiO_2 has been extensively explored by numerous research groups worldwide for a variety of applications such as photocatalytic processes for

Table 2.1. Semiconductor-based photocatalytic removal of organic dyes under visible light.

S. No.	Catalyst	Effluent	Time (min)	Degradation (%)	Reference
1	N/ZnO	Rhodamine B	20	95	[19]
2	N/TiO_2	Alizarin Red S	30	90	[20]
3	C/TiO_2	Azo dyes	60	99	[21]
4	Carbon dots/TiO_2	Methyl blue	60	98	[22]
5	$Bi@Bi_2O_3$	Rhodamine B	30	96.6	[23]
6	Fe^{3+} doped TiO_2	Malachite green	60	99	[24]
7	Bi_2O_3–ZnO	Acid black 1	60	97	[25]
8	Cr/ZnO	Azo dye	60	99	[26]
9	CdS/TiO_2	Rhodamine B	30	90	[27]
10	$CuS@CuAl_2O_4$	Eosin yellow	60	98	[28]

organic degradation, water splitting, solar cells, energy production, photocatalytic conversion, sensors, and so on. However, because TiO_2 has a large bandgap, light can only activate it [18]. Photocatalysts have been intensively investigated for many decades due to their widespread applicability in renewable energy production, organic reactions, and the abatement of environmental damage. Because of the wide range of semiconductor materials that have been synthesized, the applications of semiconductor photocatalysis have grown dramatically. As a result, attempts to find a suitable substitute have been made utilizing different metal oxides such as ZnO, MoO_3, CeO_2, ZrO_2, WO_3, α-Fe_2O_3, SnO_2, and metal chalcogenides (i.e. ZnS, CdS, CdSe, WS_2, and MoS_2), among others. However, the rapid recombination of charge carriers and the inability of O_2 to effectively trap electrons at the conduction band (CB) to produce superoxide radicals have blocked the practical use of metal oxide nanomaterials as photocatalysts. To address this problem, scientists are doping or combining pristine materials with other metal oxide materials to change the bandgap energies of the resulting composites. This reduces the recombination of charge carriers and improves the photocatalytic activity of the composites. Table 2.1 summarizes the removal of a variety of organic dyes by semiconductor metal-oxide-based photocatalysts under visible light irradiation.

The N-doped ZnO composites have demonstrated an effective photodegradation process in which approximately 95% of the RhB decomposed within 20 min. Only a ~57% decrease in RhB was seen over the same time period in control studies that used pure ZnO as the photocatalyst. Under UV light irradiation, N–ZnO nanoparticles demonstrated enhanced photocatalytic performance in comparison to pure ZnO [19]. For example, when compared to pure TiO_2, N-TiO_2 rice grains demonstrated improved UV-light-assisted photocatalysis in the photodegradation of the industrially significant anthraquinone dye alizarin red S [20]. It possessed more photocatalytic activity for the breakdown of acid black 1 (AB 1) under UV light than pure ZnO. Bi_2O_3·Bi_2O_3–ZnO is a potential choice for the photocatalytic treatment of dye effluent for 60 min, reaching an efficiency of 97% at neutral pH 7 [25].

The UV–vis absorption spectra of pure ZnO and Cr-doped ZnO both showed a bandgap absorption peak at around 365 nm. In the UV–vis absorption spectra of Cr-doped ZnO, an extra wide tail developed from 400 to 750 nm compared to that of pure ZnO. Under visible light, photocatalytic tests revealed that synthesized Cr-doped ZnO nanowires were a potential photocatalyst for the restoration of water polluted by various chemically stable azo dyes [26].

2.3 Rare-earth elements

Lanthanides, a group of elements with atomic numbers 57–71, are divided into two groups based on electron configuration [29]. The light RE elements, also known as the cerium group, include lanthanum to gadolinium (atomic numbers 57–64) and the heavy RE elements include terbium to lutetium (atomic numbers 65–71). Despite being labeled as 'rare,' these elements are abundant in the Earth's crust. The most prevalent rare-earth metals, namely cerium, lanthanum, and yttrium, have abundances (atomic fractions) comparable to those of typical industrial metals such as nickel, copper, and lead. Even though thulium is the least abundant RE, its abundance is still 200 times higher than that of gold.

The similarities between the physical and chemical characteristics of REs are strongly influenced by their electrical structure. Because *5d* and *4f* electrons have comparable energies, lanthanide elements have two sorts of electronic configuration: $4f^n\ 6s^2$ and $4f^{n-1}\ 5d^1\ 6s^2$, with $n = 1-14$. Lanthanum, cerium, gadolinium, and lutetium have electronic configurations of the $4f^{n-1}\ 5d^1\ 6s^2$ type, while the other lanthanide elements have electronic configurations of the $4f^n\ 6s^2$ type. All REs may easily lose electrons in $ns^2\ (n-1)\ d^1$ or $4f^1$, which allows them all to become M^{3+} ions. M^{2+} and M^{4+} ions can be formed by several lanthanides. In comparison to the M^{3+} state, the M^{2+} and M^{4+} states are often less stable.

2.4 The preparation of rare-earth-doped metal oxides

Lanthanide metals (Ln(III)) can drastically alter metal oxide photocatalysts' electrical, physical, and chemical characteristics. Lanthanide metal ions are useful in modifying metal oxide surfaces for diverse photocatalytic applications. Various techniques for obtaining different particle sizes, shapes, and large surface areas are being investigated by research groups (figure 2.1); these characteristics are important factors that determine the activity of the surface.

Chemical, physical, and mechanical methods such as the coprecipitation technique [30–33], the sol–gel process [34, 35], hydrothermal synthesis [36–38], the electrospinning method [39, 40], the paralysis spray method [41], the combustion method [42, 43], the sonochemical method [44, 45], microwave-assisted synthesis [46], thermal evaporation [47], etc. may all be used to prepare RE-doped metal oxides, which may be altered in terms of their sizes, forms, and other characteristics. A few of the techniques adopted and the materials prepared are detailed in table 2.2.

A study investigated the optical characteristics of ZnO nanorods codoped with Eu (1%) and Ce (1%) [53]. This study illustrated that the optical absorption and emission of ZnO are easily modified by the addition of RE ions and that the resulting

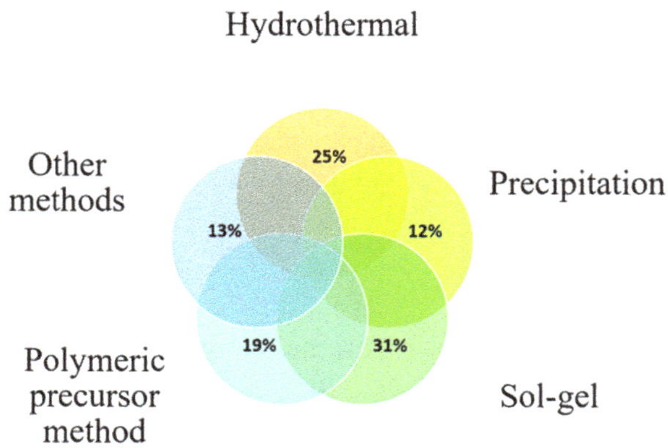

Figure 2.1. Major synthesis methods for the preparation of RE-doped metal oxides.

Table 2.2. The preparation of selected RE-doped metal oxides.

S. No.	Rare-earth-doped metal oxide	Preparation technique	Application	References
1	$LuFeO_3$	Polymeric precursor method	Photocatalysis	[48]
2	$YFeO_3$	Polymeric precursor method	Photocatalysis	[49]
3	Gd-doped CeO_2	Polymeric precursor method	Photocatalysis	[50]
4	Ce- and Y-doped SnO_2	Polymeric precursor method	Photocatalysis	[51]
5	$LaMnO_3$	Microemulsion—reverse micelle	Photocatalysis	[52]
6	Ce- and Eu-doped ZnO	Coprecipitation	Photocatalysis	[53]
7	Dy-doped ZnO	Coprecipitation	Photocatalysis	[54]
8	Ce-doped ZnO	Coprecipitation	Photocatalysis	[55]
9	Sm-doped TiO_2	Sol–gel	Photocatalysis	[56]
10	Ce-doped TiO_2	Sol–gel	Photocatalysis	[57]
11	Gd-doped SnO_2	Sol–gel	Photocatalysis	[58]
12	Eu-doped ZnO	Sol–gel	Photo- and thermoluminescence	[59]
13	Er-doped ZnO	Hydrothermal	Photoluminescence and optical properties	[60]
14	Ho-doped TiO_2	Sol–gel	Photocatalysis	[61]
15	Pr-doped molecularly imprinted TiO_2	Solvothermal	Photocatalysis	[62]
16	La-doped ZnO thin film	Dip coating	Photocatalysis	[63]

codoped material may be employed as an efficient electrode material in solar cell applications, optoelectronic devices, and photocatalysis. Under visible irradiation, Pr-doped molecularly imprinted TiO2 mesocrystals (Pr-MIP-TMCs) demonstrated higher photocatalytic activity than molecularly imprinted TiO2 mesocrystals (MIP-TMCs), which was attributed to its strong interactions between the target molecules and the imprinted cavities, its large adsorption capacity for the target molecules, its increased availability of adsorption sites, its narrow bandgap energy, and its efficient separation of photogenerated charges [62].

2.5 The photocatalytic removal of contaminants by rare-earth–metal oxides

RE-doped metal oxide photocatalysts have received a lot of interest due to their ability to split water into H_2 and O_2 when exposed to light irradiation (UV/visible/sunlight). The use of RE elements in metal oxide photocatalysts attracts a lot of interest compared to unmodified metal oxides because the *4f-levels* of lanthanide elements play a significant role in photocatalytic activities. Figure 2.2 shows the major photocatalytic applications of RE-doped metal oxides. The majority of studies have involved the photocatalytic degradation of pharmaceuticals and organic dyes. Tables 2.3–2.5 display some representative rare-earth elements used in photocatalysts for the degradation of pharmaceuticals, dyes, etc. under UV, visible, and solar irradiation.

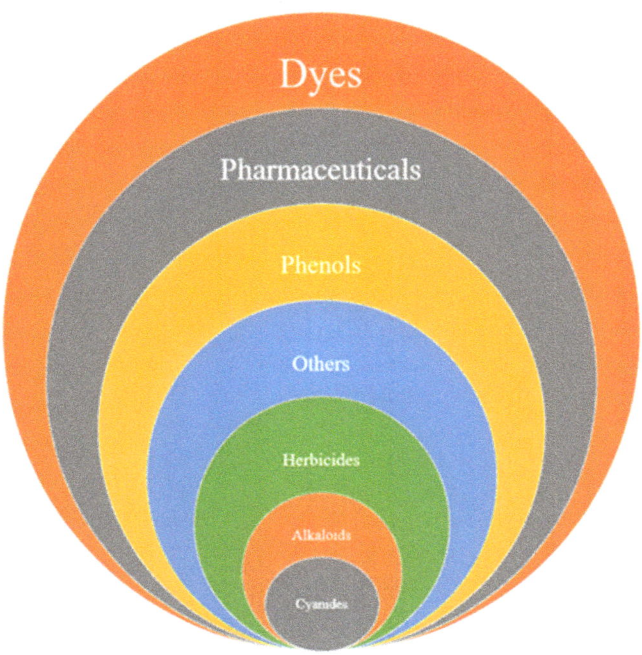

Figure 2.2. Photocatalytic applications of RE-doped metal oxides.

Table 2.3. Photocatalytic removal of endocrine disrupter.

S. No.	Catalyst	Effluent	Time (min)	Degradation (%)	References
1	Gd-doped $BiFeO_3$	Ciprofloxacin	240	80	[65]
2	FeOCl doped with Y	Ibuprofen	20	84	[66]
3	$FeCeO_x$	Diclofenac	40	83	[67]
4	$Er_2FeSbO_7/BiTiSbO_6$	Enrofloxacin	150	99.1	[68]
5	CeO_x-modified graphite felt	Carbamazepine	60	69.4	[69]
6	$LaFeO_3$/lignin biochar	Ofloxacin	75	95.6	[70]
7	CeO_2 nanosheets doped with Fe	Salicylic acid	120	96	[71]
8	$LaFeO_3$	Sulfamethoxazole	120	100	[72]
9	$LaFeO_3$/MIL-125-NH_2	Carbamazepine and caffeine	60	74 and 87	[73]
10	CeO_2/C gas diffusion electrode	Dipyrone	20	100	[74]

Table 2.4. The photocatalytic removal of organic dyes under visible light irradiation.

S. No.	Catalyst	Effluent	Time (min)	Degradation (%)	References
1	WO_3 doped with La, Gd, or Er	Methylene blue	90	98	[75]
2	$CoFe_2O_4$ doped with La or Ce	Remazol black 5, Remazol brilliant yellow	30	⩾90	[76]
3	Ce^{4+}–TiO_2 solution and nanocrystallites	Brilliant red	120	99.9	[77]
4	$PrFeO_3/CeO_2$	Methyl violet	30	80.1	[78]
5	$PrFeO_3$	Methyl violet	60		[79]
6	$LaFeO_3$ doped with Eu	Safranine-O and Remazol brilliant yellow	20	⩾90	[80]
7	$LaFeO_3$ doped with Gd		35		
8	$LaFeO_3$ doped with Dy		35		
9	$LaFeO_3$ doped with Nd		20		
10	$ZnFe_2O_4$ doped with Sm	Methylene blue	60	65	[81]
11	$ZnFe_2O_4$ doped with Dy	Methylene blue	45	97.3	[82]
12	$EuFeO_3$ nanoparticles	Rhodamine B	180	71	[83]
13	Eu^{2+} BiVO4	Methyl orange	180	93.6	[84]
14	Eu-N codoped TiO_2/ sepiolite nanocomposites	Orange G	540	98	[85]

(*Continued*)

Table 2.4. (*Continued*)

S. No.	Catalyst	Effluent	Time (min)	Degradation (%)	References
15	h-$YbFeO_3$/o-$YbFeO_3$	Methyl violet	180	64	[86]
16	Y-Bi_2WO_6	Rhodamine B	240	85	[87]
17	Yb/Er/Pr-Bi_2WO_6	Congo red	60	90.2	[88]
		Tetracycline	60	52.3	
		Rhodamine B	60	95.5	
18	Sm_2O_3/$BiVO_4$	Rhodamine B	210	55.1	[89]
	Eu_2O_3/$BiVO_4$			51.8	
	Tb_2O_3/$BiVO_4$			57.3	
19	Eu-doped Bi_2WO_6	Rhodamine B	60	98	[90]
20	Gd^{3+}-doped Bi_2MoO_6	Rhodamine B	10	84	[91]

Table 2.5. The photocatalytic removal of organic dyes under sunlight.

S. No.	Catalyst	Effluent	Time (min)	Degradation (%)	References
1	FeAl/Ce-Mts (montmorillonite)	Reactive blue 19	180	100	[92]
2	Gd-doped $BiFeO_3$	Methylene blue	240	96	[93]
		Rhodamine B	180	97	
3	Sm–ZnO	Methylene blue	90	96	[94]
4	FeAl/La-Mts	Reactive blue 19	180	100	[94]
5	Ce–CdS-G	Methylene blue	90	99	[95]
6	Gd/La@ZnO	Rhodamine B	100	91	[96]
7	Y/Ho-doped NiO	Methyl red	60	99	[97]
8	Eu–TiO_2	Methyl orange	60	98	[98]
9	NiO-CGYO (nickel-based yttria and gadolinia codoped ceria)	Methylene blue	120	90	[99]
10	$BiGdYVO_4$	Methylene blue	90	94	[100]

2.6 The photocatalytic removal of antibiotics (endocrine disrupters) by rare-earth–metal oxides

Environmental contamination has a tremendous impact on human survival and health. The overuse and misuse of antibiotics has led to a slew of issues, putting human and environmental health at risk [64]. Therefore, modern chemistry seeks excellent photocatalytic materials that can degrade antibiotics into CO_2, H_2O, and other nontoxic small molecules. Due to their particular properties, RE-doped metal oxides have emerged as next-generation photocatalysts.

Numerous approaches have been tried to remove antibiotics [65–74] from the environment using RE-doped metal oxides as catalysts. In recent years, researchers have studied the use of various RE-doped metal oxides for the degradation of antibiotics. Some degradation results are tabulated in table 2.3. One study investigated the photodegradation of the colorless organic pollutants ciprofloxacin (CIP) and levofloxacin (LFX) by a Gd-doped $BiFeO_3$ (BGFO) photocatalyst [65]. Remarkably, 240 min of sunlight was sufficient to cause deteriorations of 80% and 79% in CIP and LFX, respectively. The photocatalytic activity of BGFO may benefit from its superior shape, good crystallinity, higher optical absorption, and efficient charge carrier separation.

2.7 The photocatalytic removal of organic dye effluent by rare-earth-doped metal oxides

In photocatalysis, RE-doped metal oxide-based photocatalysts allow fast photogenerated charge carrier separation and provide abundant surface functional groups for light-harvesting materials, making their photoconversion efficiency acceptable. Therefore, numerous theoretical and experimental studies have recently shown the potential of RE-doped metal-oxide-based photocatalysts for the removal of organic dye effluents under visible light irradiation [75–91]. Several RE-doped metal-oxide-based photocatalysts have been produced and tested for their effective degradation of organic dyes (table 2.4).

WO_3 doped with La or Gd had faster and better charge separation in a composite material compared to that of pristine WO_3. The enhanced photocatalytic activity was found to occur in the following order: Gd–W > WC > Er–W > La–W > W. The findings showed excellent photocatalytic performance for 2% Gd-modified WO_3. This was attributed to the combined effects of a higher surface area, reduced charge carrier recombination, the surface hydroxyl content, the extension of photosensitivity toward the visible light absorption region, and an optimal dopant content [75].

Numerous attempts have been made to control and treat wastewater. RE-doped metal-oxide-based photocatalysts can play a significant part in the photocatalytic removal of organic dye effluents under solar irradiation [92–100]. Table 2.5 shows the photocatalytic degradation of RE-doped metal-oxide-based photocatalysts under solar irradiation. For example, methylene blue (MB) degradation under solar radiation was used to investigate the photo-induced activity of Sm-doped ZnO nanoflowers. Notably, 5% Sm–ZnO had the best degradation efficiency, achieving 96% degradation within 90 min [84].

Table 2.6 summarizes the removal of a variety of organic dyes by RE-doped metal-oxide-based photocatalysts under UV light irradiation. The photocatalytic efficiencies of ZnO nanoneedles doped with 0, 0.5, and 1% Nd were assessed by using UV light to degrade methylene blue. With 2.5 times the photocatalytic activity of undoped ZnO, the 1% Nd-doped ZnO performed best in this study [94]. As a result, this study focused on RE-doped semiconductor materials. Specifically, coprecipitation was used to synthesize cerium-doped dysprosium oxide. Under

Table 2.6.

S. No	Catalyst	Effluent	Time (Min)	Degradation %	Reference
01	ZnO-doped Nd	Methylene blue	300	92	[101]
02	Sm-doped CeO_2 nanoparticles	Rose Bengal	90	89	[102]
03	Eu^{2+}–TiO_3	Methylene blue	60	70	[103]
04	Dy-doped ZnO	Crystal violet	-	98	[104]
05	Zn–CeO_2	Methylene blue	60	97	[105]
06	Er dopant into $K_2Ta_2O_6$	Phenol	240	90	[108]
07	NiO-CGYO	Methylene blue	120	90	[99]
08	MgO- and Er^{3+}-doped MgO	Methyl Orange	250	88	[106]
09	Dy^{3} $^{+}$-doped CeO_2	Orange G	120	96	[107]
10	Eu^{3+}–TiO_3	Rhodamine B	30	96	[109]

neutral pH conditions, 10 mg of photocatalyst exhibited superior activity for 10 ppm of Oregon green (OG) dye under 96% UV light irradiation [99].

2.8 The photocatalytic mechanism of rare-earth-based photocatalysts

According to all these studies carried out by diverse researchers, RE-doped metal oxides have superior photocatalytic activity for the degradation of different organic dyes when exposed to UV, visible, and solar light. The production of radical species in the reaction media provides the primary insight into the photocatalytic performance [110]. In any semiconductor material, light irradiation excites electrons from the valence band (VB) into the conduction band (CB), and the holes present in the VB are free to react with water molecules to produce hydroxyl radicals ($^{\bullet}OH$). In addition, the superoxide radical anion (O^{2-}) is formed when oxygen in the dissolved solution reacts with the electrons in the CB. These radicals, which are powerful oxidizing species, take part in oxidation and reduction reactions. The excited electrons are held in the f subshell of the RE element(s) in the case of RE-doped metal oxides [55]. This trapping process delays the recombination process, which leads to the continuous production of reactive free radicals, which is responsible for the enhanced photocatalytic performance compared to that of the pristine metal oxide. Hence, RE-doped metal oxides show better photocatalytic activity under UV and visible light irradiation. Figure 2.3 shows the photocatalytic mechanism of RE-doped ZnO under visible light irradiation; it has been found that the photocatalyst's absorption region shifts from the UV region to the visible region upon the addition of ZnO to the Dy_2O_3 matrix. In Dy_2O_3, it has been observed that before the trapped electron returns to the ground state, h^+ reacts with the H_2O molecule to form free $^{\bullet}OH$ radicals, thereby initiating the degradation process.

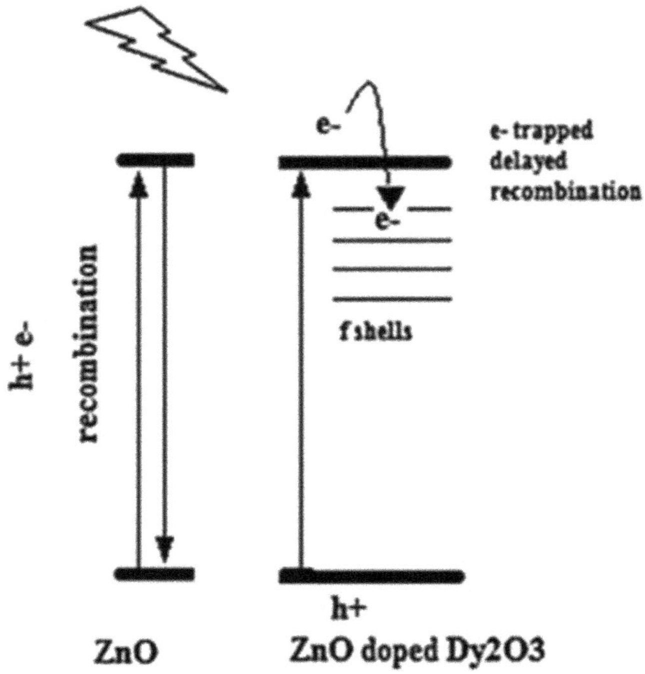

Figure 2.3. The photocatalytic mechanism of an RE-based photocatalyst. Reprinted from [55], Copyright (2014), with permission from Elsevier.

2.9 Conclusions

In brief, we conclude that RE elements are excellent dopants for metal oxides, as they enhance their properties drastically. This chapter has introduced the simple methods of synthesis employed in the research literature for the doping metal oxides with RE materials. Techniques such as the polymeric precursor method, the sol–gel method, coprecipitation, the hydrothermal method, and the solvothermal method account for the majority of the materials prepared. Since RE-doped metal oxides exhibit the characteristic *f–f transitions* associated with *F-block* elements, their properties are further enhanced. These materials contain exceptional photocatalytic, optical, and luminescent properties. Their enhanced photocatalytic property is attributed to the trapping of excited electrons in the f-*subshells*, which delays the recombination of electron–hole pairs. This further lengthens the time in which free holes can react with water molecules to generate $^{\bullet}OH$ radicals. Hence, RE-doped metal oxides serve as better nanomaterials for the degradation of organics present in wastewater under UV, visible, and natural solar irradiation. Persistent organic pollutants such as dyes, phenolics, antibiotics, and phthalates can easily be degraded. Hence, this chapter concludes that rare-earth doping of metal oxides provides better nanomaterials for environmental remediation for the treatment of wastewater under natural solar irradiation through a photocatalytic approach.

References and further reading

[1] Xiao J, Xie Y and Cao H 2015 Organic pollutants removal in wastewater by heterogeneous photocatalytic ozonation *Chemosphere* **121** 1–17
[2] 2020 *Nanoagronomy* ed S Javad (Cham: Springer Naure)
[3] Lu H, Wang J, Wang T, Wang N, Bao Y and Hao H 2017 Crystallization techniques in wastewater treatment: an overview of applications *Chemosphere* **173** 474–84
[4] Amin M T, Alazba A A and Manzoor U 2014 A review of removal of pollutants from water/wastewater using different types of nanomaterials *Adv. Mater. Sci. Eng.* **2014** 1–25
[5] Tambe Patil B B 2015 Wastewater treatment using nanoparticles *J. Adv. Chem. Eng.* **5** 10–4172
[6] Kansal S, Singh M and Sud D 2008 Studies on TiO_2/ZnO photocatalyzed degradation of lignin *J. Hazard. Mater.* **153** 412–7
[7] Li Y, Xie W, Hu X, Shen G, Zhou X, Xiang Y, Zha X and Fang P 2010 Comparison of dye photodegradation and its coupling with light-to-electricity conversion over TiO_2 and ZnO *Langmuir* **26** 591–7
[8] Tian C, Zhang Q, Wu A, Jiang M, Liang Z, Jiang B and Fu H 2012 Cost-effective large-scale synthesis of ZnO photocatalyst with excellent performance for dye photodegradation *Chem. Commun.* **48** 2858–60
[9] Fujishima A and Honda K 1972 Electrochemical photolysis of water at a semiconductor electrode *Nature* **238** 37–8
[10] Suganya G A, Josephine A and Sivasamy 2014 Nanocrystalline ZnO doped on lanthanide oxide Dy_2O_3: a novel and UV light active photocatalyst for environmental remediation *Environ. Sci. Technol. Lett.* **1** 172–8
[11] Sakthivel S, Neppolian B, Shankar M, Arabindoo B, Palanichamy M and Murugesan V 2003 Solar photocatalytic degradation of azo dye: comparison of photocatalytic efficiency of ZnO and TiO_2 *Sol. Energy Mater. Sol. Cells* **77** 65
[12] Qi K, Cheng B, Yu J and Ho W 2017 Review on the improvement of the photocatalytic and antibacterial activities of ZnO *J. Alloys Compd.* **727** 792–820
[13] Zhan W, Guo Y, Gong X, Guo Y, Wang Y and Lu G 2014 Current status and perspectives of rare earth catalytic materials and catalysis *Chinese J. Catal.* **35** 1238–50
[14] Mazierski P, Mikolajczyk A, Bajorowicz B, Malankowska A, Zaleska-Medynska A and Nadolna J 2010 The role of lanthanides in TiO_2-based photocatalysis: a review *Appl. Catal.* B **233** 301–17
[15] Yan Y and Wei S H 2008 Doping asymmetry in wide-bandgap semiconductors: origins and solutions *Phys. Status Solidi (B)* **245** 641–52
[16] Georgieva J, Valova E, Armyanov S, Philippidis N, Poulios I and Sotiropoulos S 2012 Bi-component semiconductor oxide photoanodes for the photo electrocatalytic oxidation of organic solutes and vapours: a short review with emphasis to TiO_2–WO_3 photoanodes *J. Hazard. Mater.* **211–2** 30–46
[17] Levi A, Verbitsky L, Waiskopf N and Banin U 2021 Sulfide ligands in hybrid semiconductor–metal nanocrystal photocatalysts: improved hole extraction and altered catalysis *ACS Appl. Mater. Interfaces* **14** 647–53
[18] Jiang R, Li B, Fang C and Wang J 2014 Metal/semiconductor hybrid nanostructures for plasmon-enhanced applications *Adv. Mater.* **26** 5274–309

[19] Zhang D E, Gong J Y, Ma J J, Han G Q and Tong Z W 2013 A facile method for synthesis of N-doped ZnO mesoporous nanospheres and enhanced photocatalytic activity *Dalton Trans.* **42** 16556–61
[20] Babu V J, Nair A S, Peining Z and Ramakrishna S M L 2011 Synthesis and characterization of rice grains like nitrogen-doped TiO_2 nanostructures by electrospinning–photocatalysis *Mater. Lett.* **65** 3064–8
[21] Costa E, Zamora P P and Zarbin A J G J 2011 Novel TiO_2/C nanocomposites: synthesis, characterization, and application as a photocatalyst for the degradation of organic pollutants *Colloid Interface Sci.* **368** 121–7
[22] Li F, Tian F, Liu C, Wang Z, Du Z, Li R and Zhang L 2015 One-step synthesis of nanohybrid carbon dots and TiO_2 composites with enhanced ultraviolet light active photocatalysis *RSC Adv.* **5** 8389–96
[23] Liu X, Cao H and Yin J 2011 Generation and photocatalytic activities of Bi@Bi2O3 microspheres *Nano Res.* **4** 470–82
[24] Asiltürk M, Sayılkan F and Arpaç E 2009 effect of Fe^{3+} ion doping to TiO_2 on the photocatalytic degradation of malachite green dye under UV and vis-irradiation A *J. Photochem. Photobiol. A Chem.* **203** 64–71
[25] Balachandran S and Swaminathan M 2012 Facile fabrication of heterostructured Bi_2O_3–ZnO photocatalyst and its enhanced photocatalytic activity *Phys. Chem.* C **116** 26306–12
[26] Wu C, Shen L, Zhang Y C and Huang Q 2011 Solvothermal synthesis of Cr-doped ZnO nanowires with visible light-driven photocatalytic activity *Mater. Lett.* **65** 1794–6
[27] Jiang B, Yang X, Li X, Zhang D, Zhu J and Li G J 2013 Core–shell structure CdS/TiO_2 for enhanced visible-light-driven photocatalytic organic pollutants degradation *Sol-Gel. Sci. Technol.* **66** 504–11
[28] Bellal B, Trari M and Afalfiz A 2015 Synthesis and characterization of CdS/$CuAl_2O_4$ core–shell: application to photocatalytic eosin degradation *Appl. Nanosci.* **5** 673–80
[29] Kumari M, Devi L, Maia G, Chen T, Al-Zaqri N and Ali M 2022 Mechanochemical synthesis of ternary heterojunctions TiO_2(A)/TiO_2(R)/ZnO and TiO_2(A)/TiO_2(R)/SnO_2 for effective charge separation in semiconductor photocatalysis: a comparative study *Environ. Res.* **203** 111841
[30] Pantazis D and Neese F 2009 All-electron scalar relativistic basis sets for the lanthanides *J. Chem. Theory Comput.* **5** 2229–38
[31] Tsayn C Y and Wang M C 2013 Structural and optical studies on sol–gel derived ZnO thin films by excimer laser annealing *Ceram. Int* **39** 469
[32] Raghvendra S and Pandey A 2009 and Sanjay S optical properties of europium doped bunches of ZnO nanowires synthesized by co-precipitation methods *Chalcogenide Lett.* **6** 233–9
[33] Sin J C, Lam S M, Lee K T and Mohamed A R 2013 Fabrication of erbium-doped spherical-like ZnO hierarchical nanostructures with enhanced visible light-driven photocatalytic activity *Mater. Lett.* **91** 1–4
[34] Jia T, Fu F L Z and Zhang Q 2009 Synthesis, characterization and luminescence properties of Y-doped and Tb-doped ZnO nanocrystals *Mater. Sci. Eng.* B **162** 179–84
[35] Shahroosvand H and Ghorbani-Asl M 2013 Solution-based synthetic strategies for Eu doped ZnO nanoparticle with enhanced red photoluminescence *J. Lumin.* **144** 223

[36] Wang M, Huang C, Huang Z, Guo W, Huang J, He H, Wang H, Cao Y, Liu Q and Liang J 2009 Synthesis and photoluminescence of Eu-doped ZnO microrods prepared by hydrothermal method *Opt. Mater.* **31** 1502

[37] Aneesh P M and Jayaraj M K 2010 Red luminescence from hydrothermally synthesized Eu-doped ZnO nanoparticles under visible excitation *Bull. Mater. Sci.* **33** 227

[38] Devi L S K, Kumar K S and Balakrishnan 2011 Rapid synthesis of pure and narrowly distributed Eu doped ZnO nanoparticles by solution combustion method *Mater. Lett.* **65** 35–7

[39] Dan W D, Hai Y J, Jian C, Hui L J, Ming G and Yan L X 2011 A mini-review on rare earth metal doped ZnO nanomaterials for photocatalytic remediation of waste water *Chem., Res.* **27** 174

[40] Kumar S and Sahare P D 2014 Gd^{3+} incorporated ZnO nanoparticles: a versatile material *Mater. Res. Bull.* **51** 217

[41] Liau c c and Chao L C 2010 Growth and characterization of Er doped ZnO prepared by reactive ion beam sputtering *3rd Int. Nanoelectronics Conf. (INEC) (Hong Kong)*

[42] Reddy A J, Kokila M K, Nagabhushana H, Kumara K S, Chakradhar R P S, Nagabhushana B M and Krishna R H 2014 Luminescence studies and EPR investigation of solution combustion derived Eu doped ZnO *Spectrochim. Acta, Part* A **132** 305

[43] Pessoni H V S, Maia L J Q and Franco A 2015 Eu-doped ZnO nanoparticles prepared by the combustion reaction method: Structural, photoluminescence and dielectric characterization *Mater. Sci. Semicond. Process.* **30** 135

[44] Liu J, Huang X, Li Y, Sulieman K M, Sun F and He X 2006 Selective growth and properties of zinc oxide nanostructures *Scr. Mater.* **55** 795

[45] Murmua P P, Kennedy J, Ruck B J, Markwitz A, Williams G V M and Rubanov S 2012 Structural and magnetic properties of low-energy Gd implanted ZnO single crystals *Nucl. Instrum. Methods Phys. Res.* B **272** 100

[46] Korake P V, Kadam A N and Garadkar K M 2014 Photocatalytic activity of Eu3+-doped ZnO nanorods synthesized via microwave assisted technique *J. Rare Earths* **32** 306

[47] Ishizumi A, Fujita S and Yanagi H 2011 Influence of atmosphere on photoluminescence properties of Eu-doped ZnO nanocrystals *Opt. Mater.* **33** 1116

[48] Ahmad T and Lone I H 2017 Development of multifunctional lutetium ferrite nanoparticles: structural characterization and properties *Mater. Chem. Phys.* **202** 50–5

[49] Ahmad T, Lone I H, Ansari S G, Ahmed J, Ahamad T and Alshehri S M 2017 Multifunctional properties and applications of yttrium ferrite nanoparticles prepared by citrate precursor route *Mater. Des.* **126** 331–8

[50] Martínez J M G, Meneses R A M and da Silva C R M 2014 Synthesis of gadolinium doped ceria ceramic powder by polymeric precursor method (Pechini) *Mater. Sci. Forum* **798** 182–8

[51] Fajardo H V, Longo E, Probst L, Valentini A, Carreño N, Nunes M R, Maciel A P and Leite E R 2008 Influence of rare earth doping on the structural and catalytic properties of nanostructured tin oxide *Nanoscale Res. Lett.* **3** 194–9

[52] Ahmad T, Ramanujachary K V, Lofland S E and Ganguli A K 2006 Reverse micellar synthesis and properties of nanocrystalline GMR Ramifications of size considerations *J. Chem. Sci.* **118** 513–8

[53] Murugadoss G, Jayavel R and Rajesh Kumar M 2015 Structural and optical properties of highly crystalline Ce, Eu and co-doped ZnO nanorods *Superlattices Microstruct.* **82** 538–50

[54] Suwarnkar G V K M B and Garadkar N L G K M 2016 Solgel microwave assisted synthesis of Sm-doped TiO_2 nanoparticles and their photocatalytic activity for the degradation of Methyl Orange under sunlight *J. Mater. Sci., Mater. Electron.* **27** 6425–32

[55] Suganya Josephine G A and Sivasamy A 2014 Nanocrystalline ZnO doped Dy_2O_3 a highly active visible photocatalyst: the role of characteristic f orbital's of lanthanides for visible photoactivity *Appl. Catal.* B **150– 151** 288–97

[56] Suganya Josephine G A, Jayaprakash K, Meenakshi G, Sivasamy A, Nirmala Devi G and Viswanath R N 2021 Photocatalytically active ZnO flaky nanoflowers for environmental remediation under solar light irradiation: effect of morphology on photocatalytic activity *Bull. Mater. Sci.* **44** 247

[57] Choudhury B, Borah B and Choudhury A 2012 Extending photocatalytic activity of TiO_2 nanoparticles to visible region of illumination by doping of cerium *Photochem. Photobiol.* **88** 257–64

[58] Al-hamdi A M, Sillanpää M and Dutta J 2015 Gadolinium doped tin dioxide nanoparticles: an efficient visible light active photocatalyst *J. Rare Earths* **33** 1275–83

[59] Upadhyay P K, Sharma N and Sharma S *et al* 2021 Photo and thermoluminescence of Eu doped ZnO nanophosphors *J. Mater. Sci.: Mater. Electron.* **32** 17080–93

[60] Achehboune M, Khenfouch M, Boukhoubza I, Derkaoui I, Leontie L, Carlescu A, Mothudi B M, Zorkani I and Jorio A 2022 Optimization of the luminescence and structural properties of Er-doped ZnO nanostructures: effect of dopant concentration and excitation wavelength *J. Lumin.* **246** 118843

[61] CAI H, LIU G, Lü W, LI X, YU L and LI D 2008 Effect of Ho-doping on photocatalytic activity of nanosized TiO_2 catalyst *J. Rare Earths* **26** 71–5

[62] Qi H-P and Wang H-L 2020 Facile synthesis of Pr-doped molecularly imprinted TiO_2 mesocrystals with high preferential photocatalytic degradation performance *Appl. Surf. Sci.* **511** 145607

[63] Maache A, Chergui A, Djouadi D, Benhaoua B, Chelouche A and Boudissa M 2019 Effect of La doping on ZnO thin films physical properties: correlation between strain and morphology *Optik* **180** 1018–26

[64] Shen J, Lu Y, Liu J K and Yang X H 2016 Design and preparation of easily recycled Ag_2WO_4@ZnO@Fe_3O_4 ternary nanocomposites and their highly efficient degradation of antibiotics *J. Mater. Sci.* **51** 7793–802

[65] Sharmin F and Basith M 2022 Highly efficient photocatalytic degradation of hazardous industrial and pharmaceutical pollutants using gadolinium doped $BiFeO_3$ *nanoparticles. J. Alloys Compd.* **901** 163604

[66] Fida H, Zhang G, Guo S and Naeem A 2017 Heterogeneous Fenton degradation of organic dyes in batch and fixed bed using La-Fe montmorillonite as catalyst *J. Colloid Interface Sci.* **490** 859–68

[67] Chong S, Zhang G, Zhang N, Liu Y, Zhu J, Huang T and Fang S 2016 Preparation of $FeCeO_x$ by ultrasonic impregnation method for heterogeneous Fenton degradation of diclofenac *Ultrason. Sonochem.* **32** 231–40

[68] Luan J, Liu W, Yao Y, Ma B, Niu B, Yang G and Wei Z 2022 Synthesis and property examination of Er_2FeSbO_7/$BiTiSbO_6$ heterojunction composite catalyst and light-catalyzed retrogradation of enrofloxacin in pharmaceutical wastewater under visible light irradiation *Materials* **15** 5906

[69] Wang X, Jin Y, Chen W, Zou R, Xie J, Tang Y, Li X and Li L 2021 Electro-catalytic activity of CeO_x modified graphite felt for carbamazepine degradation via E-peroxone process *Front. Environ. Sci. Eng.* **15** 122

[70] Chen X, Zhang M, Qin H, Zhou J, Shen Q, Wang K, Chen W, Liu M and Li N 2022 Synergy effect between adsorption and heterogeneous photo-Fenton-like catalysis on $LaFeO_3$/lignin-biochar composites for high efficiency degradation of ofloxacin under visible light *Sep. Purif. Technol.* **280** 119751

[71] Wang W, Zhu Q, Qin F, Dai Q and Wang X 2018 Fe doped CeO_2 nanosheets as Fenton-like heterogeneous catalysts for degradation of salicylic acid *Chem. Eng. J.* **333** 226–39

[72] Nie Y, Zhang L, Li Y-Y and Hu C 2015 Enhanced Fenton-like degradation of refractory organic compounds by surface complex formation of $LaFeO_3$ and H_2O_2 *J. Hazard. Mater.* **294** 195–200

[73] Younes H A, Taha M, Khaled R, Mahmoud H M and Abdelhameed R M 2023 Perovskite/metal-organic framework photocatalyst: a novel nominee for eco-friendly uptake of pharmaceuticals from wastewater *J. Alloys Compd.* **930** 167322

[74] Assumpção M H M T, Moraes A, De Souza R, Reis R M, Rocha R S, Gaubeur I, Calegaro M L, Hammer P, Lanza M R d V and Santos M C D 2013 Degradation of dipyrone via advanced oxidation processes using a cerium nanostructured electrocatalyst material *Appl. Catal. A Gen.* **462** 256–61

[75] Tahir M B and Sagir M 2019 Carbon nanodots and rare metals (RM = La, Gd, Er) doped tungsten oxide nanostructures for photocatalytic dyes degradation and hydrogen production *Sep. Purif. Technol.* **209** 94–102

[76] Sharma R, Bansal S and Singhal S 2016 Augmenting the catalytic activity of $CoFe_2O_4$ by substituting rare earth cations into the spinel structure *RSC Adv.* **6** 71676–91

[77] Xie Y and Yuan C 2003 Visible-light responsive cerium ion modified titania sol and nanocrystallites for X-3B dye photodegradation *Appl. Catal. B Environ.* **46** 251–9

[78] Seroglazova A S, Chebanenko M I, Nevedomskyi V N and Popkov V I 2023 Solution combustion synthesis of novel $PrFeO_3/CeO_2$ nanocomposite with enhanced photo-Fenton activity under visible light *Ceram. Int.* **49** 15468–79

[79] Seroglazova A S and Popkov V 2022 Synthesis of highly active and visible-light-driven $PrFeO_3$ photocatalyst using solution combustion approach and succinic acid as fuel *Nanosyst. Phys. Chem. Math.* **13** 649–54

[80] Dhiman M and Singhal S 2019 Effect of doping of different rare earth (europium, gadolinium, dysprosium and neodymium) metal ions on structural, optical and photocatalytic properties of $LaFeO_3$ perovskites *J. Rare Earths* **37** 1279–87

[81] Keerthana S, Yuvakkumar R, Kumar P S, Ravi G and Velauthapillai D 2021 Rare earth metal (Sm) doped zinc ferrite ($ZnFe_2O_4$) for improved photocatalytic elimination of toxic dye from aquatic system *Environ. Res.* **197** 111047

[82] Ismail M, Akhtar K, Khan M, Kamal T, Khan M A, Asiri M, Seo A and Khan J 2019 S.B. Pollution, toxicity and carcinogenicity of organic dyes and their catalytic bio-remediation *Curr. Pharm. Des.* **25** 3645–63

[83] Ju L, Chen Z, Fang L, Dong W, Zheng F and Shen M 2011 Sol–gel synthesis and photo-Fenton-like catalytic activity of $EuFeO_3$ nanoparticles *J. Am. Ceram. Soc.* **94** 3418–24

[84] Zhang A and Zhang J 2010 Effects of europium doping on the photocatalytic behavior of $BiVO_4$ *J. Hazard. Mater.* **173** 265–72

[85] Zhou F, Wang H, Zhou S, Liu Y and Yan C 2020 Fabrication of europium-nitrogen co-doped TiO_2/Sepiolite nanocomposites and its improved photocatalytic activity in real wastewater treatment *Appl. Clay Sci.* **197** 105791
[86] Tikhanova S M, Lebedev L A, Martinson K D, Chebanenko M I, Buryanenko I V, Semenov V G, Nevedomskiy V N and Popkov V I 2021 The synthesis of novel heterojunction h-$YbFeO_3$/o-$YbFeO_3$ photocatalyst with enhanced Fenton-like activity under visible-light *New J. Chem.* **45** 1541–50
[87] Cao R, Huang H, Tian N, Zhang Y, Guo Y and Zhang T 2015 Novel Y doped Bi_2WO_6 photocatalyst: hydrothermal fabrication, characterization and enhanced visible-light-driven photocatalytic activity for Rhodamine B degradation and photocurrent generation *Mater. Charact.* **101** 166–72
[88] Li X, Li W, Liu X, Geng L, Fan H, Ma X, Dong M and Qiu H 2022 The construction of Yb/Er/Pr triple-doped Bi_2WO_6 superior photocatalyst and the regulation of superoxide and hydroxyl radicals *Appl. Surf. Sci.* **592** 153311
[89] Gu S, Li W, Bian Y, Wang F, Li H and Liu X 2016 Highly-visible-light photocatalytic performance derived from a lanthanide self-redox cycle in Ln_2O_3/$BiVO_4$ (Ln: Sm, Eu, Tb) redox heterojunction *J. Phys. Chem.* C **120** 19242–51
[90] Xu X, Ge Y, Li B, Fan F and Wang F 2014 Shape evolution of Eu-doped Bi_2WO_6 and their photocatalytic properties *Mater. Res. Bull.* **59** 329–36
[91] Yu C, Wu Z, Liu R, He H, Fan W and Xue S 2016 The effects of Gd^{3+} doping on the physical structure and photocatalytic performance of Bi_2MoO_6 nanoplate crystals *J. Phys. Chem. Solids* **93** 7–13
[92] Huang Z, Wu P, Li H, Li W, Zhu Y and Zhu N 2014 Synthesis and catalytic properties of La or Ce doped hydroxy-FeAl intercalated montmorillonite used as heterogeneous photo Fenton catalysts under sunlight irradiation *RSC Adv.* **4** 6500–7
[93] Sharmin F and Basith M 2022 Highly efficient photocatalytic degradation of hazardous industrial and pharmaceutical pollutants using gadolinium doped $BiFeO_3$ nanoparticles *J. Alloys Compd.* **901** 163604
[94] Saravanan S, Patrick D S, Vangari G A, Krishna Mohan M, Ponnusamy S and Muthamizchelvan C 2022 Facile synthesis of Sm doped ZnO nanoflowers by Co-precipitation method for enhanced photocatalytic degradation of MB dye under sunlight irradiation *Ceram. Int.* **48** 29049–58
[95] Devendran P, Selvakumar D, Ramadoss G, Sivaramakrishnan R, Alagesan T, Jayavel R and Pandian K 2022 A novel visible light active rare earth doped CdS nanoparticles decorated reduced graphene oxide sheets for the degradation of cationic dye from wastewater *Chemosphere* **287** 132091
[96] Palanivel B, Macadangdang R R, Hossain M S, Alharthi F A, Kumar M, Chang J-H and Gedi S 2023 Rare earth (Gd, La) co-doped ZnO nanoflowers for direct sunlight driven photocatalytic activity *J. Rare Earths* **41** 77–84
[97] Munawar T, Fatima S and Nadeem M S *et al* 2023 Tunability of physical properties of NiO by the introduction of rare earth metal (Y, Ho) dual doping for natural sunlight-driven photocatalysis *J. Mater. Sci.: Mater. Electron.* **34** 687
[98] Khade G, Gavade N L and Suwarnkar M B *et al* 2017 Enhanced photocatalytic activity of europium doped TiO_2 under sunlight for the degradation of methyl orange *J. Mater. Sci.: Mater. Electron.* **28** 11002–11

[99] Karthik K, Radhika D, Raghava Reddy K, Raghu A V, Sadasivuni K K, Palani G and Gurushankar K 2021 Gd^{3+} and Y^{3+} co-doped mixed metal oxide nanohybrids for photocatalytic and antibacterial applications *Nano Express* **2** 010014

[100] Noor M, Sharmin F, Mamun M A A, Hasan S, Hakim M A and Basith M A 2022 Effect of Gd and Y co-doping in $BiVO_4$ photocatalyst for enhanced degradation of methylene blue dye *J. Alloys Compd.* **895** 162639

[101] Yayapao O, Thongtem T, Phuruangrat A and Thongtem S 2013 Ultrasonic-assisted synthesis of Nd-doped ZnO for photocatalysis *Mater. Lett.* **90** 83–6

[102] Keerthana S, Yuvakkumar R, Kumar P S, Ravi G and Velauthapillai D 2021 Rare earth metal (Sm) doped zinc ferrite ($ZnFe_2O_4$) forimproved photocatalytic elimination of toxic dye from aquatic system *Environ. Res.* **197** 111047

[103] Hernández J, Coste S, Murillo A G, Romo F C and Kassiba A 2017 Effects of metal doping (Cu, Ag, Eu) on the electronic and opticalbehavior of nanostructured TiO_2 *J. Alloys Compd.* **710** 355–63

[104] Dhiman P, Rana G, Kumar A, Dawi E A and Sharma G 2023 Rare Earth Doped ZnO Nanoparticles as Spintronics and Photo Catalyst for Degradation of Pollutants. *Molecules.* **28** 2838

[105] Efendi A F and Nurhasanah. I 2015 UV-light absorption and photocatalytic properties of Zn-doped CeO_2 nanopowders prepared by ultrasound irradiation *Mater. Sci. Forum* **827** 56–61

[106] Vijaya Shanthi R, Kayalvizhi R and John Abel M *et al* 2022 MgO nanoparticles with altered structural and optical properties by doping (Er^{3+}) rare earth element for improved photocatalytic activity *Appl. Phys.* A **128** 133

[107] Prathap Kumar M, Suganya Josephine G A and Sivasamy A 2017 Oxidation of organic dye using nanocrystalline rare earth metal ion doped CeO_2 under UV and visible light irradiations *J. Mol. Liq.* **242** 789–97

[108] Krukowska A, Winiarski M J, Strychalska-Nowak J, Klimczuk T, Lisowski W, Mikolajczyk A, Pinto H P, Puzyn T, Grzyb T and Zaleska-Medynska A 2018 Rare earth ions doped $K_2Ta_2O_6$ photocatalysts with enhanced UV–vis light activity *Appl. Catal.* B **224** 451–68

[109] Zhang Y, Zhang H, Xu Y and Wang Y 2003 Europium doped nanocrystalline titanium dioxide: preparation, phase transformation and photocatalytic properties *J. Mater. Chem.* **13** 2261–5

[110] Suganya Josephine G A and Sivasamy A 2014 Nanocrystalline ZnO doped on lanthanide oxide Dy_2O_3: a novel and UV light active photocatalyst for environmental remediation *Environ. Sci. Technol. Lett.* **1** 172–8

[111] 2014 *Photocatalytic Semiconductors: Synthesis, Characterization and Environmental Applications* ed A Hernández-Ramírez and I Medina-Ramírez (New York: Springer)

IOP Publishing

Photocatalysts for Energy and Environmental Sustainability

Vijay B Pawade and Bharat A Bhanvase

Chapter 3

Composite nanostructures based on metal oxide/ upconversion phosphors with improved photocatalytic activity

Kamila Zhumanova, Darya Goponenko and Timur Sh Atabaev

TiO_2, ZnO, and some other metal-oxide-based semiconductor nanostructures are traditionally employed in photocatalytic reactions such as organic pollutant degradation and hydrogen evolution. However, because of their large bandgaps, the photocatalytic activity of these nanostructures is limited; typically, these materials require highly energetic UV photons for activation. Several strategies, including doping, sensitization with plasmonic nanoparticles, light conversion with upconversion nanostructures, and control of surface porosity, have been proposed to extend the absorption range of metal-oxide-based nanostructures. In this chapter, we review recent advances in the synthesis and application of composite metal oxide/ upconversion phosphor nanostructures for photocatalytic activity enhancement. Special emphasis is placed on composites that are chemically stable, made from naturally abundant materials, reusable, and capable of being operated under solar light illumination.

3.1 Introduction

3.1.1 Metal oxide semiconductors

The electronic structure of semiconductor material is typically characterized by a filled valence band (VB) and an empty conduction band (CB) separated by a region known as the bandgap (E_g). In general, the bandgap value largely defines the optical properties of the semiconductor, specifically its absorption range. When a semiconductor is exposed to light photons with energies greater than or at least equal to the bandgap value, photoexcited electrons (e^-) move from the VB to the CB, resulting in the formation of holes (h^+) in the VB. It should be mentioned that the

doi:10.1088/978-0-7503-5697-8ch3

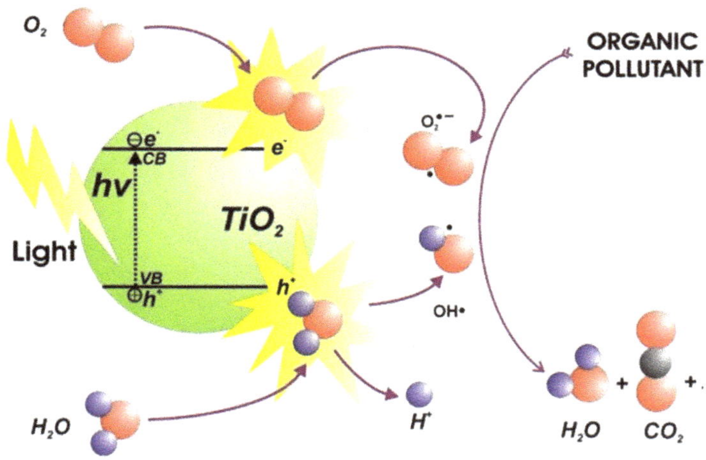

Figure 3.1. The general scheme of organic pollutant photodegradation by a TiO_2 photocatalyst. Reproduced from [3]. CC BY 4.0.

nature of the bandgap (i.e. direct or indirect bandgap) differs among various semiconductors; however, this information is outside the scope of this chapter and can be found elsewhere in the literature [1]. The photogenerated holes in the VB and the electrons in the CB can be further recombined/deactivated, react with their surroundings, or become trapped in metastable surface states. Typically, electron–hole pairs can be deactivated either radiatively in the form of light emission or non-radiatively as a release of heat. However, the deactivation process can be avoided by creating a surface defect or adding a scavenger agent that inhibits e^-/h^+ recombination. Depending on the type of scavenger, electrons or holes can react with various substances adsorbed on the semiconductor surface. For example, photogenerated holes oxidize H_2O or OH^- to produce $^\bullet OH$ radicals (equations (3.1) and (3.2)), whereas electrons produce superoxide $^\bullet O_2{}^-$ radicals, which are later transformed into $^\bullet OH$ radicals via a series of reactions (equations (3.3)– (3.6)). As powerful oxidizing agents, $^\bullet OH$ radicals cause the stepwise mineralization of organic compounds (equation (3.7)) and the formation of CO_2 and H_2O [2]. Figure 3.1 depicts the general scheme of organic pollutant photodegradation using TiO_2 (a commonly used photocatalytic semiconductor).

$$h^+ + H_2O \rightarrow {}^\bullet OH + H^+ \tag{3.1}$$

$$h^+ + OH^- \rightarrow {}^\bullet OH \tag{3.2}$$

$$e^- + O_2 \rightarrow {}^\bullet O_2{}^- \tag{3.3}$$

$${}^\bullet O_2{}^- + H^+ \rightarrow HO_2^\bullet \tag{3.4}$$

$$2HO_2^\bullet \rightarrow H_2O_2 + O_2 \tag{3.5}$$

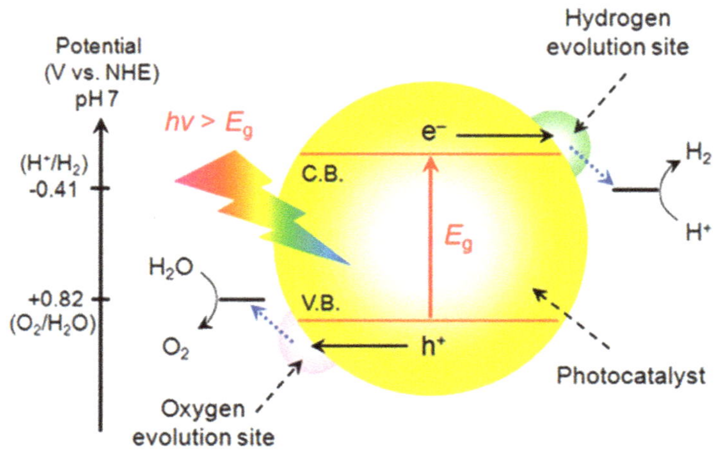

Figure 3.2. The general scheme of photochemical water splitting based on a TiO_2 photocatalyst. Reprinted with permission from [4]. Copyright (2010) American Chemical Society.

$$H_2O_2 + e^- \rightarrow {}^{\bullet}OH + OH^- \quad (3.6)$$

$$^{\bullet}OH + \text{pollutants} \rightarrow \rightarrow \rightarrow CO_2 + H_2O \quad (3.7)$$

Photoexcitation is also required for photochemical water splitting, which is a thermodynamically unfavorable process described by two half-reactions (equations (3.8) and (3.9)) versus normal hydrogen electrode (NHE) potentials.

$$2H_2O \rightarrow O_2 + 4H^+ + 4e^- \text{ (OER, oxygen evolution reaction; + 1.23 V vs. NHE)} \quad (3.8)$$

$$4H^+ + 4e^- \rightarrow 2H_2 \text{ (HER, hydrogen evolution reaction; 0 V vs. NHE)} \quad (3.9)$$

Figure 3.2 shows a typical scheme for the photochemical water splitting process. In general, several conditions must be met for photochemical water splitting to happen: (a) the photon energy should be greater than or equal to the semiconductor bandgap value; (b) the bottom level of the CB should be more negative than the reduction potential of H^+/H_2, while the top level of the VB should be more positive than the oxidation potential of H_2O/O_2; (c) the ideal semiconductor should have a suitable bandgap value (slightly higher than 1.23 eV) for effective light absorption in the visible range; (d) the semiconductor should exhibit excellent chemical stability against photocorrosion, (e) it should be made from inexpensive and naturally abundant material, and (f) appropriate cocatalysts should be used for the oxygen evolution reaction (OER) and the hydrogen evolution reaction (HER). Finding a semiconductor that meets all of these requirements is challenging, so various strategies and methods have been used to overcome all of these issues.

To date, various metal oxide semiconductors such as TiO_2, ZnO, SnO_2, $SrTiO_3$, WO_3, and others have commonly been used as photocatalytic materials because they are naturally abundant, readily available on the market, and reasonably priced. Furthermore, the majority of these metal oxide semiconductors have good chemical stability and appropriate VB and CB positions, making them suitable for both organic pollutant degradation and H_2 production. However, the light absorption properties of metal oxide semiconductors are often limited because of their large bandgap values, requiring the use of highly energetic UV–blue photons for photo-excitation. Hence, the efficiency of metal oxide semiconductors is often limited because only a small portion of the solar radiation can be utilized for catalyst activation.

The formation of Z-scheme heterostructures with spatially separated oxidation/reduction sites [5–7], doping [8], plasmonic sensitization [9, 10], coupling with optical materials [11, 12], surface engineering [13], and coupling with 2D materials [14, 15] have been shown to be efficient techniques for improving the light-harvesting ability of prepared photocatalysts. Among these, optical materials that have a frequency conversion ability (upconversion (UC) materials) are considered to be the most intriguing materials and are less frequently reviewed in the literature in combination with photocatalytic materials. Hence, in this chapter, we limit ourselves to a description and analysis of the scientific literature on UC/semiconductor systems published between 2013 and 2023, i.e. the last decade's achievements. We strongly believe that the careful literature analysis and the description of various methods in this chapter will be useful in the design/construction of complex structures with improved photocatalytic properties.

3.1.2 Upconversion materials

UC materials are spectral converters that absorb one or more low-energy photons in the near-infrared (NIR) region and emit higher-energy photons, typically in the visible spectrum, though UV photons can be observed as well. To date, UC materials have been extensively employed as spectral converters in a range of applications including photocatalysis and photovoltaics [11, 12], theranostics [16], optogenetics [17], bioimaging, disease treatment, analytical science, and biosafety [18–20]. A detailed description of UC mechanisms, such as excited-state absorption (ESA), energy transfer upconversion (ETU), cooperative sensitization upconversion (CSU), and photon avalanche (PA) can be found in the literature [16, 21]. The general structure of UC materials is a host–guest system, i.e. a host matrix doped with lanthanide ions. The selection of the host matrix is crucial for the design of highly efficient UC materials. Typically, low phonon energy host matrices are utilized to maximize radiative emission while minimizing nonradiative losses [22]. The phonon energies of chlorides, bromides, and iodides are typically less than 300 cm^{-1}, but these materials are hygroscopic and thus unsuitable for the construction of UC materials because water is a known luminescence quencher [23]. Hence, the most common host materials used to date have been fluorides, oxides, phosphates, vanadates, and so on. In particular, $NaYF_4$, which has a low phonon energy of ~300–400 cm^{-1}, is known to be the most popular and widely used host matrix [24].

Trivalent lanthanide ions such as Yb, Er, Tm, Ho, and Nd are often employed as optically active centers, and their mechanism of UC emission is based on

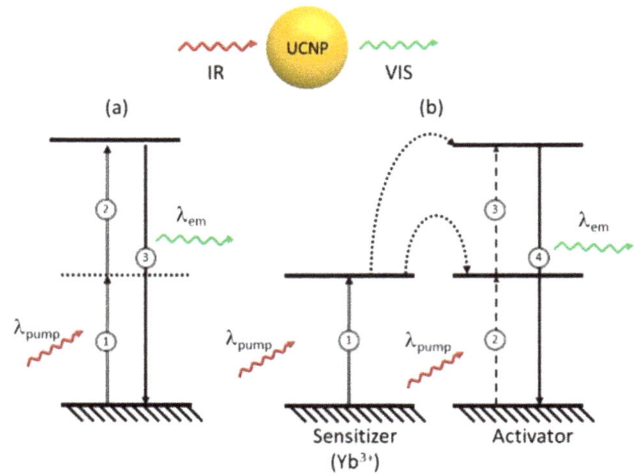

Figure 3.3. The upconversion mechanisms of the (a) ESA, and (b) ETU processes. Reproduced from [25].

crystal-field-induced intermixing of lanthanide electronic 4f states. It should be mentioned that the doping concentration of activators should be optimized to avoid the quenching of emission related to cross-relaxation effects. In the case of a single dopant (ESA), the ion should have a ladder-like multilevel system with intermediate levels that have relatively long lifetimes (figure 3.3(a)). Only a few elements can meet these requirements, so UC materials are frequently composed of two dopants, one of which acts as a sensitizer and the other as an optical activator (figure 3.3(b)). The ytterbium (Yb^{3+}) ion is frequently used as a sensitizer because its transition level $^2F_{7/2} \rightarrow {}^2F_{5/2}$ is resonant for many *f–f* transitions of other lanthanides, particularly those of Er^{3+} and Tm^{3+} ions. Moreover, the $^2F_{7/2} \rightarrow {}^2F_{5/2}$ transition of Yb^{3+} can be conveniently activated by a commercial continuous-wave (CW) 980 nm laser. Since weak *f–f* transitions couple with the local crystal field of the host matrix, the luminescence generated by a specific element is generally independent of the host material. It is also interesting to note that the emission color of a UC material can be tuned to achieve a specific color or cover a specific emission range by codoping the material with another activator. The most representative example is a codoped $NaYF_4$:Yb/Tm/Er system (figure 3.4), in which the concentrations of Yb, Tm, and Er can be varied depending on the desired color output [26].

In most cases, only highly energetic UV–blue photons can activate photocatalytic reactions because the majority of metal-oxide-based photocatalysts have large bandgaps that limit their effective absorption in the visible and NIR ranges. In this regard, Yb and Tm codoped UC systems are frequently selected for photocatalytic reactions due to the characteristic $^1I_6 \rightarrow {}^3F_4$ (~345 nm), $^1D_2 \rightarrow {}^3H_6$ (~360 nm), and $^1D_2 \rightarrow {}^3H_6$ (~450–460 nm) transitions of Tm^{3+} [27, 28]. In general, unabsorbed NIR photons are converted to UV–blue photons, which are then

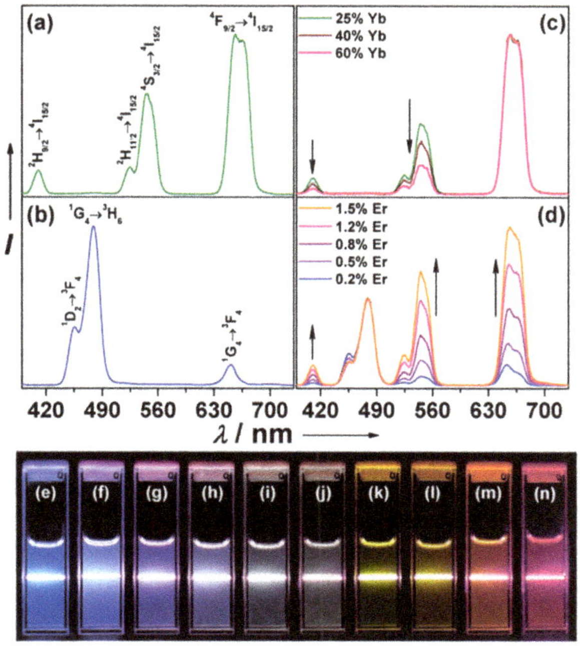

Figure 3.4. UC emissions produced by (a) $NaYF_4$:Yb^{3+},Er^{3+}, (b) $NaYF_4$:Yb^{3+},Tm^{3+}. (c) The results of varying the Yb concentration in $NaYF_4$:Yb^{3+},Er^{3+} and (d) varying the Er concentration in an $NaYF_4$:Yb/Tm/Er codoped system. (e–n) Digital images of colloidal suspensions of $NaYF_4$:Yb/Tm/Er with various Yb, Er, and Tm concentrations under 980 nm excitation. Reprinted with permission from [26]. Copyright (2008) American Chemical Society.

further absorbed by the photocatalyst. Although such a practice has been found to be feasible to boost various photocatalytic reactions, it is important to remember that the quantum yield of UC processes is low, especially for the conversion of NIR to UV–blue photons [29]. As a result, researchers are actively involved in UC emission enhancement studies using a variety of approaches, such as plasmonic enhancement [30, 31], charge compensation/codoping [32, 33], dye sensitization, surface passivation/encapsulation with inorganic materials [34, 35], and energy management [36]. Typically, Yb^{3+} and Tm^{3+}-based UC materials with greatly improved UC luminescence are useful for improving a variety of photocatalytic reactions.

3.2 Composite nanostructures based on metal oxide/upconversion phosphors

Generally, large bandgaps, poor conductivity, and fast recombination of photo-generated electron–hole pairs significantly reduce the photocatalytic activity of metal oxides. To broaden the absorption range of metal oxide nanostructures in the visible region, various techniques such as metal/nonmetal doping, plasmonic sensitization, and coupling with other semiconductors have been extensively employed. Regardless of the methods mentioned above, there is still room for improvement in solar light capture. For example, UC materials have the potential to convert low-energy photons in the NIR region into high-energy UV–visible photons, which can then be reabsorbed by the semiconductor to generate more electron–hole pairs. Indeed, there are many research studies in the literature that report the fabrication of UC-semiconductor hybrid structures for photocatalytic reactions. Although such structures also demonstrate high potential in nanomedicine and various chemical transformations, we limit ourselves to a discussion of the UC metal-oxide-based hybrids reported in the last ten years (2013–23) for photocatalytic water treatment and green hydrogen generation.

3.2.1 Upconversion-capable photocatalytic materials for water treatment

The core–shell structure is the most common and rational design for UC-semiconductor hybrids because all the photons unabsorbed by the semiconductor reach and excite the UC structure, resulting in emission in the visible region that can be further absorbed by the semiconductor for electron–hole pair generation (figure 3.5). Close contact between the UC material and the semiconductor, as well as direct energy transfer, can efficiently extend solar light utilization and improve photocatalytic activity.

Typically, various UC-semiconductor structures have been prepared using the core–shell design. The most important parameters such as structure, pollutant model, degradation efficiency, and illumination source for some representative examples are listed in table 3.1. As can be seen, UC-semiconductor structures have been widely employed to degrade a variety of textile dyes, antibiotics, and industrial pollutants.

In some cases, a combination of two semiconductors can also be beneficial in terms of photocatalytic activity improvement. For example, Guo and coauthors [45] decorated $NaYF_4$:Yb^{3+},Tm^{3+} (NYT) rods with CdS and TiO_2 nanoparticles to investigate the effect of the two additional semiconductors on the photocatalytic degradation of methylene blue (MB) dye. Morphological and chemical analyses

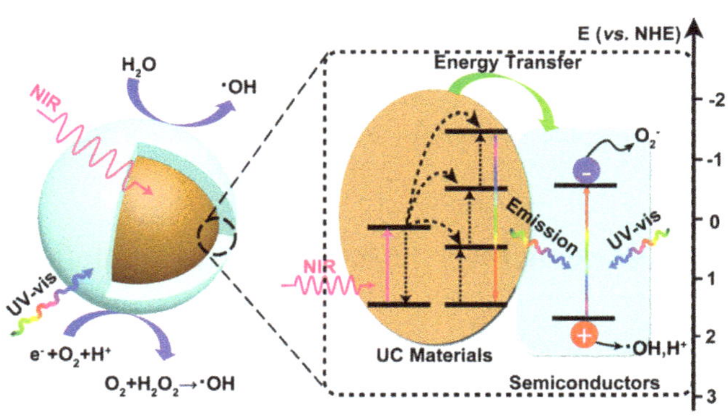

Figure 3.5. A plausible mechanism for energy transfer between a UC material and a semiconductor in a core–shell structure. Reproduced from [37] with permission from The Royal Society of Chemistry.

Table 3.1. Representative examples of UC-semiconductor structures with a core–shell design.

Structure	Pollutant (volume)	Degradation efficiency (pollutant concentration)	Illumination source	References
$NaYF_4$:Yb^{3+},Tm^{3+}@ZnO	Rhodamine B (RhB) dye (0.5 ml)	65% in 30 h (20 mg l^{-1})	980 nm diode laser (2.0 W cm^{-2})	[38]
$NaYF_4$:Yb^{3+},Tm^{3+}@TiO_2	Ciprofloxacin (volume information is not available)	99% in 2 h for optimal sample (1×10^{-5} M)	Xenon lamp (200 W)	[39]
$NaYF_4$:Yb^{3+},Tm^{3+}@TiO_2	Deoxynivalenol (3 ml)	95.7% in 60 min (10 μg ml^{-1})	Xenon lamp (50 W)	[40]
$NaYF_4$@CdS	Methyl orange (MO) dye, carbendazim (CZ) (25 ml for both pollutants)	52% in 6 h for MO and 64% in 6 h for CZ (5×10^{-5} M for both pollutants)	NIR light from solar simulator (OAI Trisol, AM 1.5, 100 mW cm^{-2})	[41]
$NaYF_4$:Yb^{3+},Tm^{3+}@ZnO	Rhodamine B (RhB) dye (50 ml)	95% in 60 min (5 mg l^{-1})	Xenon lamp (1500 mW cm^{-2})	[42]
$NaYF_4$:Yb^{3+},Tm^{3+}@ZnO	Methylene blue (MB) dye (0.5 ml)	68.7% in 10 h (15 mg l^{-1})	980 nm diode laser (10 W cm^{-2})	[43]
MoS_2-$NaYF_4$:Yb^{3+},Er^{3+}	Rhodamine B (RhB) dye (10 ml)	61% in 12 h (25 mg l^{-1})	980 nm diode laser (detailed information is not available)	[44]

(figure 3.6) revealed that hexagonal NYT rods with smooth surfaces were coated with a porous layer of CdS and TiO_2 nanoparticles. Photocatalytic experiments at 980 nm showed that the degradation rate constant for NYT/CdS/TiO_2 was ~3 and ~6.5 times higher than those of bare NYT/CdS and NYT/TiO_2, respectively. It has been shown that the synergy of two semiconductors, CdS and TiO_2, plays an important role in the separation of charge carriers and results in increased photocatalytic activity. Another study [46] also found that the degradation rate constant for a $NaYF_4$:Yb,Tm@TiO_2/CdS composite structure was ~1.6 times higher than that of bare $NaYF_4$:Yb,Tm@TiO_2 when illuminated by a visible light (200–1000 nm) source.

Some research studies have focused on the development of recyclable composite structures that can be continuously reused several times. In this context, the use of a magnetic field to quickly remove photocatalytic composites can be facilitated by the co-addition of magnetic particles. As demonstrated in figure 3.7, UC structures ($NaYF_4$:Yb,Tm) can be placed on a magnetic core and then coated with a TiO_2 shell layer [47]. The photocatalytic activity of a prepared magnetic composite and other

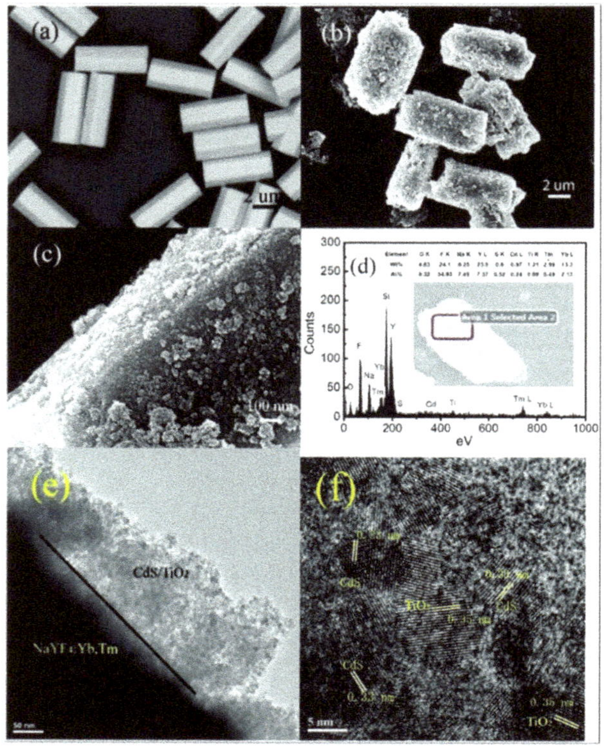

Figure 3.6. SEM images of NYT (a) and NYT/CdS/TiO_2 (b and c). Energy dispersive x-ray (EDX) spectroscopic analysis of NYT/CdS/TiO_2 (d). Transmission electron microscopy (TEM) images of NYT/CdS/TiO_2 (e and f) at various magnifications. Reproduced from [45] with permission from The Royal Society of Chemistry.

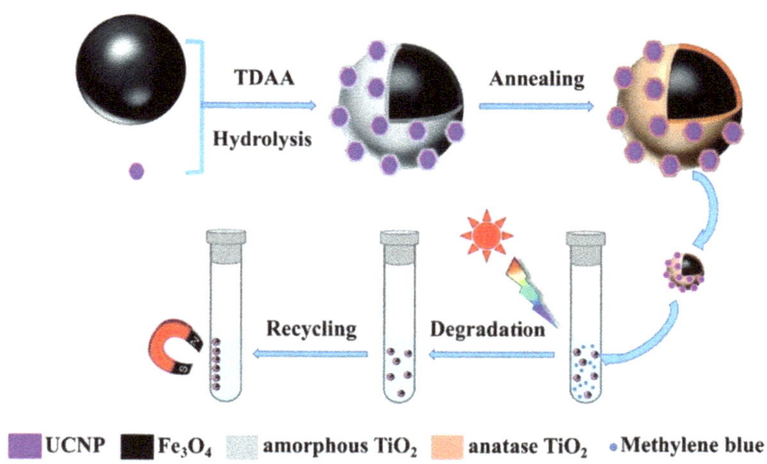

Figure 3.7. The preparation steps used for the synthesis of the magnetically recoverable (Fe_3O_4-UC)@TiO_2 composite. Reproduced from [47] with permission from The Royal Society of Chemistry.

reference samples was assessed by degrading MB dye under UV, UV–vis–NIR, and NIR illumination using mercury and xenon lamps with appropriate light filters. The authors showed that the magnetically recoverable (Fe_3O_4-UC)@TiO_2 composite exhibited higher photocatalytic activity under NIR and UV–vis–NIR illumination than commercial P25 TiO_2 nanoparticles. The reusability of the (Fe_3O_4-UC)@TiO_2 composite was tested four times and no significant loss in photocatalytic activity was found. Chen *et al* [48] utilized another design by decorating hexagonal prisms of $NaYF_4$:Yb,Tm with magnetic Fe_3O_4@SiO_2 and different amounts of P25 TiO_2 nanoparticles. Photocatalytic activity tests under NIR illumination revealed that the optimal sample, denoted by FS/UC/T20, was several times more active in MB, RhB, methyl orange, and phenol degradation than P25 TiO_2 nanoparticles. Recyclability tests, on the other hand, revealed that the photocatalytic activity dropped to ~5%–10% after the fourth cycle, which can be attributed to the partial loss of TiO_2 from the surface.

Plasmonic metal nanostructures are frequently coupled to semiconductor structures for efficient charge separation and light absorption in the visible region [49, 50]. In this regard, the mutual use of plasmonic and UC materials provides an undeniable advantage in terms of light absorption extension for semiconductor structures. For example, Ma and coauthors [51] tested a core–shell $NaYF_4$:Yb, Tm@TiO_2 composite decorated with Ag nanoparticles for R6G dye degradation under 50 W xenon lamp illumination. It was found that the degradation rate constant for $NaYF_4$:Yb,Tm@TiO_2/Ag was more than twice that of a $NaYF_4$:Yb, Tm@TiO_2 structure and ~3 times higher than that of bare TiO_2. Xu and colleagues [52] produced $NaYF_4$:Yb,Er,Tm microspheres that were coated with a porous TiO_2 layer and then decorated with varying amounts of Au nanoparticles. Figure 3.8 depicts the preparation strategy along with accompanying TEM images demonstrating the distinct morphological differences between the prepared UC structure, TiO_2-coated, and Au-decorated composite structures. Optimization experiments revealed that among all the tested samples, NYF@TiO_2–1 wt% Au had the highest degradation rate of MO dye under UV, visible, IR, and simulated solar light irradiation. In particular, the degradation rate of NYF@TiO_2–1 wt% Au nanoparticles under solar light was found to be nearly twice that of P25 TiO_2 nanoparticles. Experiments using radical scavengers revealed that both $^{\bullet}OH$ and $^{\bullet}O_2^-$ radicals were engaged in MO photodegradation; however, the contribution of the $^{\bullet}O_2^-$ radicals was more substantial, which can be attributed to effective charge separation and the supply of additional electrons by plasmonic Au nanoparticles. A similar construction might be employed to deposit other semiconductor and metal nanoparticles on the surface of an UC structure. For example, Tian *et al* [53] prepared hexagonal $NaYF_4$:Yb,Tm nanoplates coated with a photocatalytic SnO_2 shell layer and decorated with Ag nanoparticles. The samples were subsequently evaluated for RhB dye degradation under UV, Vis, NIR, and solar radiation. The degradation rates of $NaYF_4$@SnO_2@Ag were reported to be ~11.45 times (Vis), ~5.25 times (NIR), and ~2.29 times higher than that of bare $NaYF_4$@SnO_2 under solar illumination. Finite-difference time-domain (FDTD) calculations revealed a considerable electromagnetic boost at the junction between the SnO_2 and the Ag,

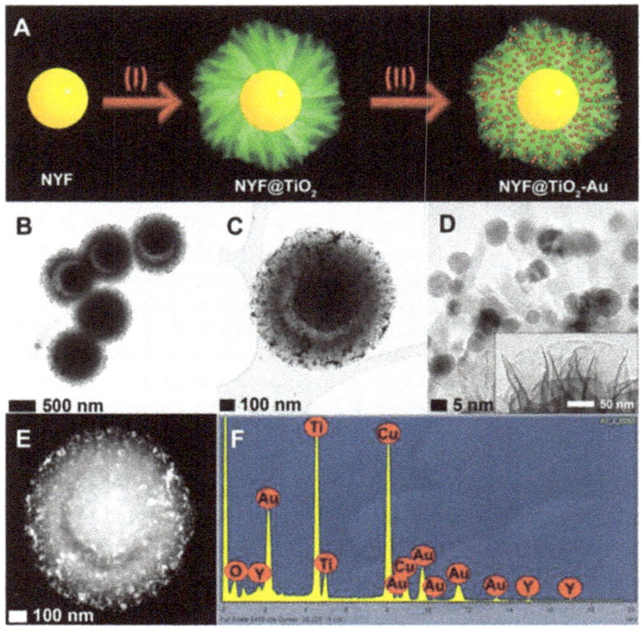

Figure 3.8. The synthesis of $NaYF_4$:Yb,Er,Tm microspheres coated with porous TiO_2 and decorated with Au nanoparticles (A). A TEM image of the prepared $NaYF_4$:Yb,Er,Tm microspheres (B); $NaYF_4$:Yb,Er,Tm coated with porous TiO_2 (C); and the final composite decorated with Au nanoparticles (D). A scanning TEM image (E) and EDX (F) analysis of the final composite. Reproduced from [52] John Wiley & Sons. © 2015 WILEY-VCH Verlag GmbH & Co. KGaA, Weinheim.

which was caused by localized surface plasmon resonance effects. As a result, the generation rate of electron–hole pairs increased dramatically, which in turn improved the photocatalytic activity of the $NaYF_4$@SnO_2@Ag composite.

3.2.2 Upconversion-capable photocatalytic materials for hydrogen generation

In general, UC-containing materials for hydrogen generation follow the same design principles as those of composites for wastewater treatment. The primary distinction is that these composites are often decorated with metallic cocatalysts such as Pt, Au, and Ag, which are used to lower the activation energy of H_2O oxidation and H^+ reduction; they also provide plasmonic sensitization and function as electron sinks for effective charge separation. Among them, Pt is the most active element for hydrogen evolution according to the Gibbs free energy of H-absorption (ΔG_H) calculations, and is hence mostly used for photocatalytic hydrogen generation [54]. Nevertheless, Au- and Ag-decorated structures are also frequently employed. For example, Feng and colleagues [55] examined the photocatalytic hydrogen evolution of a $NaYF_4$:Yb,Er/Au/CdS architecture with a CdS semiconductor as the outer shell layer. Controlled studies with no catalyst, $NaYF_4$:Yb,Er (NYF), NYF/Au, NYF/CdS, NYF/Au/CdS, and NYF/Au/CdS-M (mechanically mixed NYF and Au/CdS)

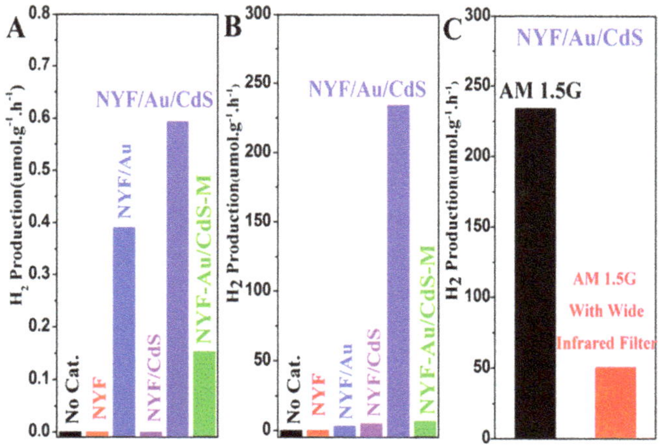

Figure 3.9. The photocatalytic activity of different samples under (A) NIR, (B) solar light, and (C) solar light without an NIR component. Reproduced from [55] with permission from The Royal Society of Chemistry.

samples were performed under NIR (A), simulated solar light (B), and solar light without NIR (C), as shown in figure 3.9. One can observe that all samples containing the NYF component are highly active under NIR illumination, emphasizing the crucial role of UC structures in photocatalytic activity improvement. Furthermore, samples containing Au also demonstrated improved catalytic performance under both NIR and solar light, indicating the necessity of cocatalyst use. The increased photocatalytic activity of the $NaYF_4$:Yb,Er/Au/CdS composite can be explained by several factors, including (a) additional photon harvesting by the UC structure, (b) plasmonic sensitization of UC and CdS structures, (c) charge carrier separation by Au, and (d) the formation of an intimate contact between NYF/Au and CdS during solvothermal synthesis.

A similar core–shell design was used in the creation of Er^{3+}:$Y_3Al_5O_{12}$/Pt–TiO_2 [56] and Er^{3+}:$YAlO_3$/Pt–TiO_2 [57] composites. Despite the fact that the resulting H_2 evolution rates were not as outstanding as in recent reports, photocatalytic improvement was evident in both cases with the inclusion of UC structures. For example, a recent study found that $NaYF_4$:Yb,Er particles coated with a TiO_2 layer and decorated with 1% Au nanoparticles generated ~350 μmol h^{-1} of hydrogen under simulated solar light irradiation [58]. This value was ~61 times and ~3.2 times greater than those of bare $NaYF_4$:Yb,Er particles covered with a TiO_2 layer and a reference TiO_2 (P25)/1% Au catalyst, respectively. Hence, the beneficial role of Au cocatalyst loading and $NaYF_4$:Yb,Er utilization is clearly apparent.

In some cases, a semiconductor may serve as the host matrix for the construction of UC structures. For example, Shang and colleagues [59] created an Er-Yb:TiO_2/MoO_{3-x} heterostructure in which the Er-Yb:TiO_2 was a two-in-one UC and photocatalytic structure, while the MoO_{3-x} compound acted as a light absorption material that generated 'hot electrons.' The results indicated that the Er-Yb:TiO_2/MoO_{3-x} composite was active under both NIR and solar light, and the H_2

generation yield from ammonia borane under simulated sunlight was ~2.5 and ~52 times higher than those of MoO_{3-x} and bare ammonia borane, respectively. Liu and coworkers [60] proposed a Z-scheme WO_3:Yb^{3+},Er^{3+}/Ag/Ag_3VO_4/Ag photocatalyst suitable both for solar antibiotic degradation and hydrogen evolution. In this case, the WO_3:Yb^{3+},Er^{3+} acted as a photocatalyst as well as a UC material, supplying additional photons to plasmonic Ag nanoparticles. It was shown that such a Z-scheme configuration can yield ~489 μmol g^{-1} of H_2 in 180 min and that it can effectively degrade organic pollutants under solar illumination.

Solar hydrogen generation can also be carried out via the photoelectrochemical water splitting method, which utilizes the same charge separation process as particle-based hydrogen generation. In this scenario, a thin-film semiconductor is coupled to a metallic electrode where hydrogen is normally evolved. Nanostructuring, doping, heterojunction formation, optical/plasmonic sensitization, and control of defects are common approaches used with semiconductor thin films to improve the generated photocurrent density that is commonly measured and compared to 1.23 V vs. RHE (the minimum potential difference for water electrolysis vs. the reversible hydrogen electrode). It is interesting that even the simple placement of UC structures on the surface of semiconductor thin films can significantly improve the generated photocurrent density. For example, the photocurrent density was increased by ~16% when Pt-doped hematite (α-Fe_2O_3) nanorods were decorated with ZrO_2:Er^{3+} nanocrystals [61]. In a similar manner, the photocurrent density measured at 1.23 V vs. RHE was increased to ~34% by drop-casting $NaYF_4$-based UC nanoparticles on the surface of a $ZnFe_2O_4$/TiO_2 photoanode [62]. Further investigation found that the UC-modified $ZnFe_2O_4$/TiO_2 photoanode demonstrated good activity under IR illumination, suggesting the beneficial role of UC nanostructures in photocurrent improvement. Furthermore, the idea of combining multiple materials, such as porous semiconductor films with ordered structures, with metallic and UC nanostructures appears to be promising in terms of maximum light collection in the UV, visible, and NIR regions. This concept can be realized using an inverse-opal structure, such as TiO_2, as shown in figure 3.10. Two approaches for UC and Au nanoparticle deposition were employed and various UC structures were tested, such as $NaYF_4$:Yb,Er and $NaYF_4$:Yb,Tm [63]. A porous structure with an ordered structure generally improves light scattering effects and, as a result, TiO_2 absorbance in the visible range. Furthermore, an increased specific area due to a porous structure can be also advantageous for photocatalytic H_2 production. It is important to note that H_2 evolution differed between the $NaYF_4$:Yb,Er and $NaYF_4$:Yb,Tm structures. For example, an optimized sample of $NaYF_4$:Yb,Er yielded ~14.1 μmol cm^{-2}, while a sample of $NaYF_4$:Yb,Tm yielded ~10.9 μmol cm^{-2} after 5 h of sunlight illumination. A detailed analysis suggested that green emission from $NaYF_4$:Yb, Er sensitizes Au nanoparticles, whereas blue emission from $NaYF_4$:Yb,Tm primarily sensitizes TiO_2 films. Hence, this study found that the sensitization of the plasmonic structure was found to be more advantageous than the direct sensitization of the semiconductor structure.

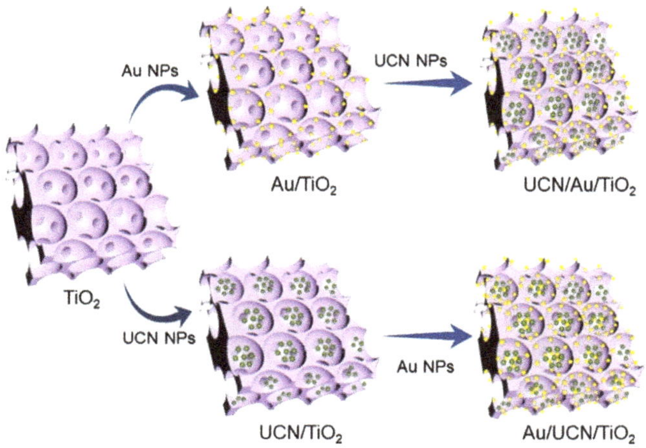

Figure 3.10. Two preparation pathways for an inverse-opal TiO_2 film loaded with an UC material and Au nanoparticles. Reprinted with permission from [63]. Copyright (2019) American Chemical Society.

3.3 Conclusions and future prospects

It is apparent that UC materials play an important role in various processes by supplying additional high-energy photons converted from NIR photons. Countless studies available in the research literature have demonstrated that the use of UC materials can improve any photocatalytic process. Although this feature is extremely useful in many applications, it is still uncontrollable, since photocatalytic process enhancement is primarily associated with the quality of the UC material, its emission pattern, its position in the composite structure, size, shape, and so on. Hence, it is strongly recommended that the quantum yields (QYs) of UC structures should be measured to make the methods of researchers working in different parts of the world somewhat comparable. Typically, thorough QY measurement approaches for UC materials can easily be found in the literature [64].

Another critical issue that can be addressed is the low QY of UC materials. Furthermore, weak UC luminescence can quickly be quenched by interactions with polar solvents. Hence, proper crystal engineering, surface passivation, and plasmonic sensitization could be key steps in improving the QY of UC materials. As previously mentioned, the selection of the UC material and even its placement in the composite structure may play key roles in photocatalytic processes. Therefore, careful design of the photocatalytic structures should be carried out in future studies. Moreover, these composites may be combined with 2D materials to accelerate charge carrier transfer and separation. Nevertheless, solar-harnessing devices with UC structures have the potential to be widely employed in the future green economy. Typically, they can be employed in conjunction with solar cells and photocatalytic cells for green H_2 generation and solar wastewater treatment.

Acknowledgments

This research was funded by the Science Committee of the Ministry of Science and Higher Education of the Republic of Kazakhstan (Grant No. AP19676347).

References and further reading

[1] Yuan L D, Deng H X, Li S S, Wei S H and Luo J W 2018 Unified theory of direct or indirect band-gap nature of conventional semiconductors *Phys. Rev.* B **98** 245203

[2] Nursam N M, Wang X and Caruso R A 2015 High-throughput synthesis and screening of titania-based photocatalysts *ACS Comb. Sci.* **17** 548–69

[3] Ibhadon A O and Fitzpatrick P 2013 Heterogeneous photocatalysis: recent advances and applications *Catalysts* **3** 189–218

[4] Maeda K and Domen K 2010 Photocatalytic water splitting: recent progress and future challenges *J. Phys. Chem. Lett.* **1** 2655–61

[5] Maeda K 2013 Z-scheme water splitting using two different semiconductor photocatalysts *ACS Catal.* **3** 1486–503

[6] Xu K, Zhang L, Yu J, Wageh S, Al-Ghamdi A A and Jaroniec M 2018 Direct Z-scheme photocatalysts: principles, synthesis, and applications *Mater. Today* **21** 1042–63

[7] Li J, Yuan H, Zhang W, Jin B, Feng Q, Huang J and Jiao Z 2022 Advances in Z-scheme semiconductor photocatalysts for the photoelectrochemical applications: a review *Carbon Energy* **4** 294–331

[8] Medhi R, Marquez M D and Lee T R 2020 Visible-light-active doped metal oxide nanoparticles: review of their synthesis, properties, and applications *ACS Appl. Nano Mater.* **3** 6156–85

[9] Atabaev T S 2018 Plasmon-enhanced solar water splitting with metal oxide nanostructures: a brief overview of recent trends *Front. Mater. Sci.* **12** 207–13

[10] Amirjani A, Amlashi N B and Ahmadiani Z S 2023 Plasmon-enhanced photocatalysis based on plasmonic nanoparticles for energy and environmental solutions: a review *ACS Appl. Nano Mater.* **6** 9085–123

[11] Yang W, Li X, Chi D, Zhang H and Liu X 2014 Lanthanide-doped upconversion materials: emerging applications for photovoltaics and photocatalysis *Nanotechnology* **25** 482001

[12] Atabaev T S and Molkenova A 2019 Upconversion optical nanomaterials applied for photocatalysis and photovoltaics: recent advances and perspectives *Front. Mater. Sci.* **13** 335–41

[13] Bai S, Jiang W, Li Z and Xiong Y 2015 Surface and interface engineering in photocatalysis *ChemNanoMat* **1** 223–39

[14] Luo B, Liu G and Wang L 2016 Recent advances in 2D materials for photocatalysis *Nanoscale* **8** 6904–20

[15] Shanker G S, Biswas A and Ogale S 2021 2D materials and their heterostructures for photocatalytic water splitting and conversion of CO_2 to value chemicals and fuels *J. Phys.: Energy* **3** 022003

[16] Chen G, Qiu H, Prasad P N and Chen X 2014 Upconversion nanoparticles: design, nanochemistry, and applications in theranostics *Chem. Rev.* **114** 5161–214

[17] Patel M, Meenu M, Pandey J K, Kumar P and Patel R 2022 Recent development in upconversion nanoparticles and their application in optogenetics: a review *J. Rare Earths* **40** 847–61

[18] Gulzar A, Xu J, Yang P, He F and Xu L 2017 Upconversion processes: versatile biological applications and biosafety *Nanoscale* **9** 12248–82
[19] Liang G, Wang H, Shi H, Wang H, Zhu M, Jing A, Li J and Li G 2020 Recent progress in the development of upconversion nanomaterials in bioimaging and disease treatment *J. Nanobiotechnol.* **18** 154
[20] Borse S, Rafique R, Murthy Z V P, Park T J and Kailasa S K 2022 Applications of upconversion nanoparticles in analytical and biomedical sciences: a review *Analyst* **147** 3155–79
[21] Auzel F 2004 Upconversion and anti-stokes processes with *f* and *d* ions in solids *Chem. Rev.* **104** 139–74
[22] Zhou J, Liu Q, Feng W, Sun Y and Li F 2015 Upconversion luminescent materials: advances and applications *Chem. Rev.* **115** 395–465
[23] Arppe R, Hyppänen I, Perälä N, Peltomaa R, Kaiser M, Würth C, Christ S, Resch-Genger U, Schäferling M and Soukka T 2015 Quenching of the upconversion luminescence of $NaYF_4:Yb^{3+},Er^{3+}$ and $NaYF_4:Yb^{3+},Tm^{3+}$ nanophosphors by water: the role of the sensitizer Yb^{3+} in non-radiative relaxation *Nanoscale* **7** 11746–57
[24] Rabouw F T, Prins P T and Norris D J 2016 Europium-doped $NaYF_4$ nanocrystals as probes for the electric and magnetic local density of optical states throughout the visible spectral range *Nano Lett.* **16** 7254–60
[25] Naccache R, Martín Rodríguez E, Bogdan N, Sanz-Rodríguez F, de la Cruz M C I, de la Fuente Á J, Vetrone F, Jaque D, Solé J G and Capobianco J A 2012 High resolution fluorescence imaging of cancers using lanthanide ion-doped upconverting nanocrystals *Cancers* **4** 1067–105
[26] Wang F and Liu X 2008 Upconversion multicolor fine-tuning: visible to near-infrared emission from lanthanide-doped $NaYF_4$ nanoparticles *J. Am. Chem. Soc.* **130** 5642–3
[27] Chen G, Qiu H, Fan R, Hao S, Tan S, Yang C and Han G 2012 Lanthanide-doped ultrasmall yttrium fluoride nanoparticles with enhanced multicolor upconversion photoluminescence *J. Mater. Chem.* **22** 20190–6
[28] Li X, Zhang F and Zhao D 2015 Lab on upconversion nanoparticles: optical properties and applications engineering via designed nanostructure *Chem. Soc. Rev.* **44** 1346–78
[29] Meijer M S *et al* 2018 Absolute upconversion quantum yields of blue-emitting $LiYF_4:Yb^{3+},Tm^{3+}$ upconverting nanoparticles *Phys. Chem. Chem. Phys.* **20** 22556–62
[30] Wu D M, Garcia-Etxarri A, Salleo A and Dionne J A 2014 Plasmon-enhanced upconversion *J. Phys. Chem. Lett.* **5** 4020–31
[31] Dong J, Gao W, Han Q, Wang Y, Qi J, Yan X and Sun M 2019 Plasmon-enhanced upconversion photoluminescence: mechanism and application *Rev. Phys.* **4** 100026
[32] Atabaev T S, Piao Z, Hwang Y H, Kim H K and Hong N H 2013 Bifunctional $Gd_2O_3:Er^{3+}$ particles with enhanced visible upconversion luminescence *J. Alloys Compd.* **572** 113–7
[33] Sinha S, Mahata M K, Swart H C, Kumar A and Kumar K 2017 Enhancement of upconversion, temperature sensing and cathodoluminescence in the K^+/Na^+ compensated $CaMoO_4:Er^{3+}/Yb^{3+}$ nanophosphor *New J. Chem.* **41** 5362–72
[34] Ti Z, Liang T, Wang Q and Liu Z 2020 Strategies for constructing upconversion luminescence nanoprobes to improve signal contrast *Small* **16** 1905084
[35] Zhu X, Zhang J, Liu J and Zhang Y 2019 Recent progress of rare-earth doped upconversion nanoparticles: synthesis, optimization, and applications *Adv. Sci.* **6** 1901358

[36] Sheng T, Xu M, Li Q, Wu Y, Zhang J, Liu J, Zhu X and Zhang Y 2021 Elucidating the role of energy management in making brighter, and more colorful upconversion nanoparticles *Mater. Today Phys.* **20** 100451
[37] Tian Q, Yao W, Wu W and Jiang C 2018 NIR light-activated upconversion semiconductor photocatalysts *Nanoscale Horiz.* **4** 10–25
[38] Guo X, Song W, Chen C, Di W and Qin W 2013 Near-infrared photocatalysis of β-$NaYF_4$:Yb^{3+},Tm^{3+}@ZnO composites *Phys. Chem. Chem. Phys.* **15** 14681–8
[39] Ma Y and Li S 2019 $NaYF_4$:Yb,Tm@TiO_2 core@shell structures for optimal photocatalytic degradation of ciprofloxacin in the aquatic environment *RSC Adv.* **9** 33519–24
[40] Wu S, Wang F, Li Q, Wang J, Zhou Y, Duan N, Niazi S and Wang Z 2020 Photocatalysis and degradation products identification of deoxynivalenol in wheat using upconversion nanoparticles@TiO_2 composite *Food Chem.* **323** 126823
[41] Balaji R, Kumar S, Reddy K L, Sharma V, Bhattacharyya K and Krishnan V 2017 Near-infrared driven photocatalytic performance of lanthanide-doped $NaYF_4$@CdS core–shell nanostructures with enhanced upconversion properties *J. Alloys Compd.* **724** 481–91
[42] Zhang F, Wang W-N, Cong H-P, Luo L-B, Zha Z-B and Qian H-S 2017 Facile synthesis of upconverting nanoparticles/zinc oxide core–shell nanostructures with large lattice mismatch for infrared triggered photocatalysis *Part. Part. Syst. Charact.* **34** 1600222
[43] Zhang J, Zhao S, Xu Z, Zhang L, Zuo P and Wu Q 2019 Near-infrared light-driven photocatalytic $NaYF_4$:Yb,Tm@ZnO core/shell nanomaterials and their performance *RSC Adv.* **9** 3688–92
[44] Chatti M, Adusumalli V N K B, Ganguli S and Mahalingam V 2016 Near-infrared light triggered superior photocatalytic activity from MoS_2–$NaYF_4$:Yb^{3+}/Er^{3+} nanocomposites *Dalton Trans.* **45** 12384–92
[45] Guo X, Di W, Chen C, Liu C, Wang X and Qin W 2014 Enhanced near-infrared photocatalysis of $NaYF_4$:Yb, Tm/CdS/TiO_2 composites *Dalton Trans.* **43** 1048–54
[46] Xu J, Shi Y, Chen Y, Wang Q, Cheng J and Li P 2018 Enhanced photocatalytic activity of TiO_2 in visible and infrared light through the synergistic effect of upconversion nanocrystals and quantum dots *Mater. Res. Express* **6** 025055
[47] Lv Y, Yue L, Li Q, Shao B, Zhao S, Wang H, Wu S and Wang Z 2018 Recyclable (Fe_3O_4-$NaYF_4$:Yb,Tm)@TiO_2 nanocomposites with near-infrared enhanced photocatalytic activity *Dalton Trans.* **47** 1666–73
[48] Chen Z and Fu M L 2018 Recyclable magnetic Fe_3O_4@SiO_2/β-$NaYF_4$:Yb^{3+},Tm^{3+}/TiO^2 composites with NIR enhanced photocatalytic activity *Mater. Res. Bull.* **107** 194–203
[49] Birnal P, Marco de Lucas M C, Pochard I, Herbst F, Heintz O, Saviot L, Domenichini B and Imhoff L 2023 Visible-light photocatalytic degradation of dyes by TiO_2–Au inverse opal films synthesized by atomic layer deposition *Appl. Surf. Sci.* **609** 155213
[50] Em S, Yedigenov M, Khamkhash L, Atabaev S, Molkenova A, Poulopoulos S G and Atabaev T S 2022 Uncovering the role of surface-attached Ag nanoparticles in photodegradation improvement of Rhodamine B by ZnO-Ag nanorods *Nanomaterials* **12** 2882
[51] Ma Y, Liu H, Han Z, Yang L and Liu J 2015 Non-ultraviolet photocatalytic kinetics of $NaYF_4$:Yb,Tm@TiO_2/Ag core@comby shell nanostructures *J. Mater. Chem.* A **3** 14642–50
[52] Xu Z, Quintanilla M, Vetrone F, Govorov A O, Chaker M and Ma D 2015 Harvesting lost photons: plasmon and upconversion enhanced broadband photocatalytic activity in core@-shell microspheres based on lanthanide-doped $NaYF_4$, TiO_2, and Au *Adv. Funct. Mater.* **25** 2950–60

[53] Tian Q, Yao W, Wu W, Liu J, Wu Z, Liu L, Dai Z and Jiang C 2017 Efficient UV–Vis–NIR responsive upconversion and plasmonic-enhanced photocatalyst based on lanthanide-doped $NaYF_4/SnO_2/Ag$ *ACS Sustain. Chem. Eng.* **5** 10889–99
[54] Zhu J, Hu L, Zhao P, Lee L Y S and Wong K Y 2020 Recent advances in electrocatalytic hydrogen evolution using nanoparticles *Chem. Rev.* **120** 851–918
[55] Feng W, Zhang L, Zhang Y, Yang Y, Fang Z, Wang B, Zhang S and Liu P 2017 Near-infrared-activated $NaYF_4:Yb^{3+}$, Er^{3+}/Au/CdS for H_2 production via photoreforming of bio-ethanol: plasmonic Au as light nanoantenna, energy relay, electron sink and co-catalyst *J. Mater. Chem.* A **5** 10311–20
[56] Li Y, Guo Y, Li S, Li Y and Wang J 2015 Efficient visible-light photocatalytic hydrogen evolution over platinum supported titanium dioxide nanocomposites coating up-conversion luminescence agent ($Er^{3+}:Y_3Al_5O_{12}/Pt–TiO_2$) *Int. J. Hydrogen Energy* **40** 2132–40
[57] Ma C, Li Y, Zhang H, Chen Y, Lu C and Wang J 2015 Photocatalytic hydrogen evolution with simultaneous photocatalytic reforming of biomass by $Er^{3+}:YAlO_3/Pt–TiO_2$ membranes under visible light driving *Chem. Eng. J.* **273** 277–85
[58] Chilkalwar A A and Rayalu S S 2018 Synergistic plasmonic and upconversion effect of the (Yb,Er)NYF-TiO_2/Au composite for photocatalytic hydrogen generation *J. Phys. Chem.* C **122** 26307–14
[59] Shang J, Xu X, Liu K, Bao Y, Yangyang and He M 2019 LSPR-driven upconversion enhancement and photocatalytic H_2 evolution for Er–Yb:TiO_2/MoO_{3-x} nano-semiconductor heterostructure *Ceram. Int.* **45** 16625–30
[60] Liu Z, Chen L, Piao C, Tang J, Liu Y, Lin Y, Fang D and Wang J 2021 Highly active Z-scheme $WO_3:Yb^{3+},Er^{3+}/Ag/Ag_3VO_4/Ag$ photocatalyst with efficient charge transfer and light utilization for enhanced levofloxacin degradation with synchronous hydrogen evolution *Appl. Catal. A.* **623** 118295
[61] Atabaev T S, Vu H H T, Ajmal M, Kim H K and Hwang Y H 2015 Dual-mode spectral convertors as a simple approach for the enhancement of hematite's solar water splitting efficiency *Appl. Phys.* A **119** 1373–7
[62] Lim Y, Lee S Y, Kim D, Han M K, Han H S, Kang S H, Kim J K, Sim U and Park Y I 2022 Expanded solar absorption spectrum to improve photoelectrochemical oxygen evolution reaction: synergistic effect of upconversion nanoparticles and $ZnFe_2O_4/TiO_2$ *Chem. Eng. J.* **438** 135503
[63] Boppella R, Mota F M, Lim J W, Kochuveedu S T, Ahn S, Lee J, Kawaguchi D, Tanaka K and Kim D H 2019 Plasmon and upconversion mediated broadband spectral response in TiO_2 inverse opal photocatalysts for enhanced photoelectrochemical water splitting *ACS Appl. Energy Mater.* **2** 3780–90
[64] Jones C M S, Gakamsky A and Marques-Hueso J 2021 The upconversion quantum yield (UCQY): a review to standardize the measurement methodology, improve comparability, and define efficiency standards *Sci. Technol. Adv. Mater.* **22** 810–48
[65] Würth C, Fischer S, Grauel B, Alivisatos A P and Resch-Genger U 2018 Quantum yields, surface quenching, and passivation efficiency for ultrasmall core/shell upconverting nanoparticles *J. Am. Chem. Soc.* **140** 4922–8

Chapter 4

Nanocomposites and their applications in photocatalytic degradation processes

Shubham Bonde, Gauri Kallawar, Bharat A Bhanvase and Bhaskar Sathe

4.1 Introduction

In today's world, the needs for clean water and large-scale food production are becoming increasingly important due to the growing global population and heightened industrial demands. However, the use of agrochemicals to enhance agricultural productivity has led to concerns regarding their adverse impacts on the environment and human health. Pesticides are known to have properties such as low biodegradability, high bioaccumulation, and long persistence, which can cause significant risks to soil, water, and food chains. In addition, wastewater dye effluents from various industries have emerged as major water pollutants, posing threats to aquatic ecosystems and human health. Researchers are actively exploring technological and non-technological approaches to address the urgent need for wastewater treatment. Various adsorptive and advanced oxidation processes (AOPs) such as the Fenton method [1], the photo-Fenton method [2], heterogeneous photocatalysis [3], and ozonation [4] systems have been studied for their ability to remove and degrade organic pollutants. AOPs utilize potent free radicals such as hydroxyl ($HO^{\bullet}$) and superoxide ($O_2^{-\bullet}$) radicals with strong oxidizing capabilities to break down pollutants into lower-molecular-weight intermediates and inorganic precursors. Heterogeneous photocatalysis has the potential to be used as an effective AOP wherein a semiconductor is activated by sunlight or artificial light. It utilizes water molecules and dissolved oxygen in oxidation–reduction reactions to efficiently degrade organic pollutants, including dyes, drugs, and pesticides. Commonly used metallic nano oxides such as zinc oxide [5] and titanium dioxide [6] have favorable properties such as low toxicity, chemical stability, large surface area/porosity, and photocorrosion resistance. However, they face drawbacks such as high bandgap energy, charge recombination, and agglomeration during photocatalysis [7]. To address these issues, researchers are exploring the use of nanocomposites, in which

doi:10.1088/978-0-7503-5697-8ch4

the active substance is dispersed in low concentrations (around 0.5–5 wt%) on a support. This approach mitigates the drawbacks associated with conventional nanocatalysts and enhances the overall efficiency of the photocatalytic process for environmental remediation.

This chapter delves into the catalytic applications of nanomaterials, particularly nanocomposites, in the treatment of pharmaceutical wastewater, pesticides, organic dyes, insecticides, soil and air. The high surface area of nanocomposites directly influences their catalytic roles, and nanomaterials have proven valuable in achieving high selectivity, specificity, stability, and recovery due to their nanoscale nature. Various types of nanomaterials, including nanofibers, nanoparticles, and nanoplatelets, have demonstrated their unique roles in catalysis. Hybrid nanomaterials that combine the distinct properties of two components have shown great promise. These nanocomposites, ranging from organic/organic to inorganic/organic and polymer/inorganic composites, offer enhanced properties that can be leveraged to catalyze the degradation of pollutant effluents. The catalytic efficacy of nanosized semiconductor photocatalysts, such as ZnO, TiO_2, and SiO_2, has been extensively explored, providing insights into their application for wastewater treatment.

This chapter also examines how graphene and altered graphene demonstrate catalytic capabilities when paired with different inorganic compounds such as metals, metal oxides, and polymers. The following sections discuss selected studies and progress in the realm of nanocomposites, highlighting their involvement in catalyzing the breakdown of dye effluents and their contribution to ongoing initiatives aimed at addressing water pollution.

4.2 Nanocomposite materials

Nanocomposite materials, considered a crucial category of advanced materials, are solid multiphase materials in which one phase possesses at least one dimension that is less than 100 nm. This process entails integrating nanoparticles into larger matrices, such as polymers, metals, ceramics, or composites. This incorporation imparts exceptional properties and enhanced performance characteristics to nanocomposites, making them valuable for various applications. At the nanoscale, they exhibit unique features, such as an increased surface-to-volume ratio, which significantly enhances catalytic efficiency by providing a higher density of active sites for adsorption—often doubling it compared to the active site densities of bulk materials [8]. As discussed in the previous section, nanocomposites address limitations associated with nanoparticle-based photocatalytic activity.

Nanocomposite photocatalysts fall into different types: metal-, polymer-, ceramic-, and carbon-based. Each type contributes significantly to environmental purification. Metal nanocomposite photocatalysts boost pollutant degradation by leveraging the combined impact of metal nanoparticles and semiconductors [9]. They efficiently enhance the separation of electron–hole pairs, speeding up the process of photodegradation. This makes them crucial for purifying water [10] and air [11]. Polymer nanocomposite photocatalysts are advanced materials used for environmental remediation. They consist of nanoscale photocatalytic particles embedded in polymer matrices. These materials efficiently degrade pollutants under UV or visible light, addressing critical pollution issues [12]. Ceramic nanocomposite photocatalysts employ nanostructures within ceramic materials to enhance the light-driven degradation of pollutants, making them effective for environmental cleanup. These materials possess a large surface area, enhanced photocatalytic activity, and the capability to efficiently degrade both organic and inorganic pollutants in air and water [13]. Carbon-based nanocomposites blend carbon structures with photocatalytic nanoparticles, improving pollutant degradation under light exposure, and showing potential for environmental remediation and wastewater treatment [14].

Nanocomposites exhibit promise as materials with improved photocatalytic activity, which is attributed to their better light absorption, enhanced charge separation, and increased surface area. Various techniques are employed to comprehend their features. X-ray diffraction (XRD) and transmission electron microscopy (TEM) are utilized to analyze their crystal structure, particle size, and morphology. Meanwhile, X-ray photoelectron spectroscopy (XPS) and energy-dispersive X-ray spectroscopy (EDS) help to identify chemical composition and elemental distribution. UV–Vis spectroscopy is applied to assess optical properties, focusing on absorption and bandgap features. Brunauer–Emmett–Teller (BET) analysis is employed to measure surface area and porosity, crucial factors influencing catalytic activity. Photoluminescence (PL) and transient absorption spectroscopy techniques are used to probe photogenerated charge separation. In addition, scanning electron microscopy (SEM) aids in visualizing surface morphology and particle size, providing insights into the nanocomposite's physical structure and uniformity.

4.2.1 Types of nanocomposite materials

Nanocomposite photocatalysts are broadly categorized into metal-, carbon- polymer-, and ceramic-based nanocomposites. Metal oxides, such as titanium dioxide (TiO_2) and zinc oxide (ZnO), display excellent photocatalytic activity under UV light, which can be enhanced by combining them with other nanoscale materials. Metal sulfides such as cadmium sulfide (CdS) and zinc sulfide (ZnS) are effective in the visible light range. Carbon-based nanocomposites involve graphene, a single-layer carbon lattice, and carbon nanotubes (CNTs), which improve charge separation and transport. Polymer-based nanocomposites include stable combinations of polymers (e.g. polyvinyl alcohol) with metal oxides, which offer improved mechanical properties, and conjugated polymers (e.g. polythiophene), known for their visible light absorption and enhanced charge separation. These nanocomposites stand at the forefront of advancements in sustainable technologies. Further, ceramic-based nanocomposites are being considered because of their inherent chemical stability, expansive surface area which effectively enables pollutant adsorption, and catalytic activity, particularly in the case of advanced oxidation processes.

4.2.1.1 Metal-based nanocomposites

Metal-based nanocomposite photocatalysts have become increasingly important in environmental remediation. These materials exhibit superior photocatalytic activity and unique properties. They are produced by incorporating metal nanoparticles into different support matrices, such as carbon nanotubes [15], metal oxides [16], or graphene [17], which work synergistically to enhance their efficiency in degrading diverse environmental pollutants. The use of metal-based nanocomposites is a viable and sustainable method with which to tackle challenges associated with air and water pollution, fostering the creation of environmentally friendly technologies for the elimination of pollutants. Precise control over their structure, morphology, and composition through rational design and synthesis allows their performance to be optimized for targeted pollutant removal, overcoming the limitations of traditional nanoparticles. For example, traditional photocatalysts such as TiO_2 nanoparticles work primarily using electron–hole recombination within particles, hindering chemical transformations. Their substantial bandgap of 3.2 eV requires the use of UV light for activation, resulting in inefficient solar energy utilization (only 5% of the spectrum) [18, 19]. Titania's nonporous, polar surface limits the absorption of nonpolar organic contaminants. Agglomeration and aggregation adversely impact TiO_2 nanoparticle photoactivity and light absorption. It is critical to address these issues to enhance the effectiveness of TiO_2 in photocatalytic applications. To overcome these constraints, nanocomposites have been utilized to augment the light absorption capacity of titanium dioxide (TiO_2) and broaden its spectral responsiveness to encompass the visible spectrum. This approach alters the photocatalyst's properties, reducing electron–hole recombination and introducing structural defects that trap electronic carriers, thereby extending their lifespans. For instance, a ZnO/TiO_2 nanocomposite photocatalyst with an 18–42 nm particle size was synthesized by Upadhyay *et al* [20] using the technique known as the sol–gel process. The experimental outcomes demonstrated that the introduction of 4% ZnO resulted in a notable increase in the degradation of methylene blue (MB), elevating the efficiency of the degradation process from 75% to 90%. Furthermore, the incorporation of ZnO led to enhanced absorption of visible light during the experiment. Bethi *et al* [21] synthesized nano photocatalysts by incorporating cerium and iron into TiO_2 using ultrasound. The application of ultrasound induced cavitation during the synthesis process. The resulting photocatalysts were then utilized in the decolorization of crystal violet dye. Notably, in the investigation of crystal violet dye decolorization, it was observed that TiO_2 doped with 0.8% iron exhibited the highest level of photocatalytic activity. An investigation conducted by Sukriti *et al* [22] reported the successful fabrication of nanoparticles comprising samarium-doped zinc oxide. The study demonstrated notable photocatalytic efficacy in terms of the maximum photodegradation of MB dye, particularly with the 4% samarium-doped photocatalyst, achieving an efficiency of 94.94% under visible light irradiation. The efficiency of the photocatalyst in degrading MB exhibited a substantial augmentation, reaching 6.98 times that of its undoped zinc oxide counterpart. $CdS–TiO_2$ nanocomposites prepared by Qutub *et al* [23] demonstrated heightened photocatalytic performance in the presence of visible light. The CdS–TiO2 nanocomposites demonstrated an 84% degradation efficiency of the organic dye AB-29, surpassing the efficiencies of CdS or TiO_2 alone. The strategy involved utilizing the UV region for TiO_2 excitation and the visible region for CdS excitation, optimizing electron–hole pair generation and trapping. In another study by Karimi-Maleh *et al* [24], metal oxide/Ag nanocomposites were successfully synthesized and achieved highly efficient photocatalytic degradation of organic dyes. The TiO_2/Ag nanocomposite, characterized by a bandgap of 2.78 eV and a remarkably increased surface area of 68.2 $m^2.g^{-1}$ compared to that of ZnO/Ag, exhibited superior photocatalytic performance. Both nanocomposites demonstrated efficient degradation of methyl orange (MO) and 4-chlorophenol under visible light, and TiO_2/Ag surpassed expectations by degrading 86% of MO within 180 min. In conclusion, maximizing the potential of advanced metal-based nanocomposite photocatalysts requires further research to optimize cost-effective and scalable synthesis processes, ensure long-term stability and reusability, and address environmental and health concerns. Collaboration between researchers, engineers, and policymakers is vital to transforming polluted areas into thriving ecosystems.

4.2.1.2 Polymer-based nanocomposites

Polymer-based nanocomposite photocatalysts have emerged as promising materials for environmental remediation, particularly in photocatalytic pollutant degradation. These photocatalysts, combining polymeric matrices such as polyvinyl alcohol (PVA) with nanoparticles such as titanium dioxide (TiO_2), exhibit enhanced photocatalytic performance [25]. The synergy between the polymer matrix and nanomaterials enables efficient charge separation and migration, improving pollutant degradation under light irradiation. Tunable properties, including the surface area and bandgap, allow for customized design, optimizing efficiency for specific pollutant degradation. The polymer matrix ensures stability and recyclability, making the materials durable and economically viable for repeated photocatalysis cycles. Some nanocomposites incorporate biocompatible polymer matrices, enhancing eco-friendliness for water treatment applications. These materials find use in water and air purification systems, as well as self-cleaning coatings for diverse surfaces, showcasing versatility in degrading organic dyes, pesticides, and pharmaceuticals and addressing emerging contaminants. Additionally, certain nanomaterials extend absorption to visible light, enabling efficient pollutant degradation under natural sunlight. In a recent investigation conducted by Malesic-Eleftheriadou *et al* [26], PET/TiO_2 nanocomposite films were fabricated with varying TiO_2 contents (10, 30, and 47 wt%) using the phase-inversion method. These films demonstrated photocatalytic activity in the degradation of a mixture of antibiotics under simulated solar irradiation. Initially, heightened TiO_2 concentrations contributed to improved photocatalytic efficiency. However, PET-30 wt% TiO_2 and PET-47 wt% TiO_2 exhibited instability in subsequent cycles, as active oxidative species generated in the initial cycle attacked PET chains, resulting in TiO_2 particle leaching. In contrast, PET-10 wt% TiO_2 maintained stability even after five cycles, effectively degrading antibiotics such as lincomycin, moxifloxacin, isoniazid, metronidazole, and norfloxacin within 2 h. Extended reaction times (6 h) were required for other compounds, which reached degradation levels of up to 98%. In another study, Feizpoor *et al* [27] synthesized TiO_2/$CoMoO_4$/polyaniline (PANI) ternary photocatalysts using a refluxing technique that incorporated $CoMoO_4$ and PANI into TiO_2. The resulting nanocomposites displayed augmented BET-specific surface areas: 45.8 $m^2\ g^{-1}$ for TiO_2, 50.1 $m^2\ g^{-1}$ for TiO_2/$CoMoO_4$ (30%), and 57.9 $m^2\ g^{-1}$

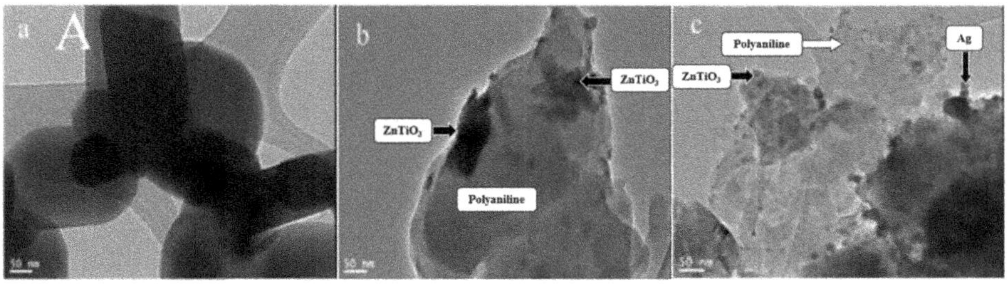

Figure 4.1. TEM images depicting (a) $ZnTiO_3$, (b) 10%PANI/$ZnTiO_3$, and (c) 1%Ag/10%PANI/$ZnTiO_3$ nanocomposite photocatalysts. Reprinted from [28], Copyright (2021), with permission from Elsevier.

for TiO_2/$CoMoO_4$/PANI (30%). The heightened surface area of TiO_2/$CoMoO_4$/PANI (30%) contributed to enhanced visible-light photocatalytic performance. In comparison to TiO_2 and TiO_2/$CoMoO_4$, TiO_2/$CoMoO_4$/PANI (30%) exhibited notable efficiency in the degradation of organic pollutants and reduction of Cr(VI). For instance, the efficiency of RhB photodegradation performed using TiO_2/$CoMoO_4$/PANI (30%) surpassed that of pure TiO_2 by 21.7 times and that of TiO_2/$CoMoO_4$ (30%) by 4.9 times. Moreover, the efficiency of the nanocomposite in Cr(VI) reduction exceeded those of TiO_2 and TiO_2/$CoMoO_4$ (30%) by 13.8 and 4.9 times, respectively. Similarly, Faisal *et al* [28] successfully synthesized Ag/PANI/$ZnTiO_3$ nanocomposites through a cost-effective sol–gel method followed by ultrasonication. XRD analysis disclosed the presence of the rhombohedral phase in $ZnTiO_3$. Figure 4.1 shows that Ag nanoparticles (10–20 nm) were attached to $ZnTiO_3$ nanoparticles (~70 nm), along with sheet-like PANI particles (>300 nm). The ternary photocatalyst, denoted by 1%Ag/10%PANI/$ZnTiO_3$, showcased outstanding performance, achieving a notable 95.6% removal of the target MB molecule. This degradation rate surpassed that of undoped $ZnTiO_3$ by 1.85 times. The heightened performance was ascribed to the diminutive size of the Ag nanoparticles and the synergistic interactions between the Ag, PANI, and $ZnTiO_3$ constituents. Furthermore, the ternary framework exhibited notable efficacy in the visible-light degradation of the insecticide imidacloprid.

Kumar *et al* [29] successfully employed an in-situ codeposition and oxidation approach to synthesize nanocomposites featuring both rutile and anatase phases. The resulting catalysts exhibited a spherical morphology. Their bandgap energies, determined using Tauc plots, were found to be 3.2, 2.86, 2.6, and 2.3 eV for TiO_2, pure PANI, TiO_2/PANI (T/P), and TiO_2/PANI/GO (T/P/G), respectively. The surface areas of TiO_2, T/P, and T/P/G were measured and found to be 39.2, 78.8, and 94.4 $m^2\ g^{-1}$, respectively. Photocatalytic investigations under visible light using T/P/G, T/P, and TiO_2 at neutral pH with a catalyst dose of 1600 mg l^{-1} demonstrated significant photodegradation (85%–99%, 60%–97%, and 10%–20%) of thymol blue (TB) and rose bengal (RB) at a concentration of 25 ppm over 180 min. Reductions in total organic carbon (TOC) substantiated photodegradation, and a kinetic study revealed first-order kinetics. Figure 4.2 visually depicts the enhanced nanocomposite-driven rate of dye photodegradation.

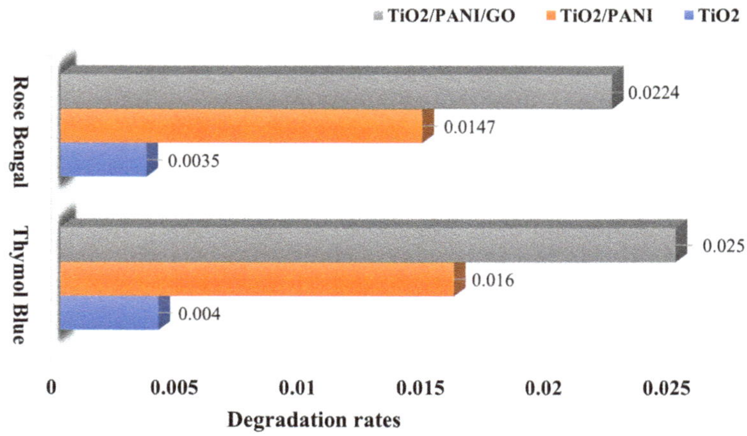

Figure 4.2. The photodegradation rates of thymol blue and rose bengal achieved using various nanocomposites. Drawn based on data from [29] Copyright (2023) from Elsevier.

Nano-fillers possess the capacity to increase the overall functionality of nanocomposite photocatalysts. This increment is achieved through an enhancement in the surface area, an enhancement in the pollutant adsorption capabilities, and an acceleration of reaction rates. Functionalized nano-fillers can selectively target pollutants, and they also reinforce strength, enhance stability, and increase resistance. The impact of these fillers on the photocatalytic activity of polymer-based photocatalysts depends on factors such as their type, dispersion, and functionalization [30]. When coupled with efficient photocatalytic activity, these fillers are invaluable in environmental remediation. Further research and development in this field could lead to sustainable solutions for addressing contemporary pollution challenges.

4.2.1.3 Ceramic-based nanocomposites

In the area of environmental remediation, ceramic-based nanocomposites are the preferred choice due to their inherent chemical stability, expansive surface area that facilitates effective pollutant adsorption, and catalytic activity, particularly in the context of AOPs. Their durability ensures sustained effectiveness in harsh conditions, which is crucial for long-term applications such as water treatment [31]. Notably, ceramics such as titanium dioxide exhibit photocatalytic properties and are capable of degrading pollutants when exposed to light, as demonstrated in photocatalytic water treatment. The diverse range of ceramic materials offers flexibility in tailoring nanocomposites to target specific pollutant types and environmental conditions. Cerium oxide, known for its resistance to chemical corrosion and UV absorption in the 385–400 nm range [32] has proved to be an effective catalyst. Doping CeO_2 ceramics with suitable dopants enhances their photocatalytic performance [33]. An innovative methodology presented by Zinatloo-Ajabshir *et al* [34] employed a novel environmentally friendly reducing and stabilizing agent—Phoenix dactylifera extract. This extract played a crucial role in the synthesis of a nanostructure composed of $ZnCo_2O_4$ and Co_3O_4. This nanomaterial exhibited remarkable efficiency under visible light, achieving energy gap values of approximately 2.68 and 2.87 eV for the $ZnCo_2O_4$–Co_3O_4 product and pure $ZnCo_2O_4$, respectively, in nanoscale particles ranging from 30 to 55 nm. The photocatalytic degradation efficiencies for RB, acid black 1, and phenol red reached 100%, 95.7%, and 100%, respectively, after 75 min of exposure to visible light. The $ZnCo_2O_4$–Co_3O_4 nanostructure had excellent stability and recyclability over nine consecutive runs. In another study, Mahdavi *et al* [35] synthesized novel photocatalytic nanocomposites (Dy_2O_3-SiO_2) utilizing tetraethylenepentamine (Tetrene) as a basic agent through a facile and efficient sonochemical method. Morphological analyses revealed that optimal control of the sonication time (10 min) and ultrasonic power (400 W) resulted in the formation of a porous nanocomposite comprising spherical nanoparticles with sizes ranging from 20 to 60 nm. The binary Dy_2O_3-SiO_2 nanophotocatalyst exhibited an energy gap of 3.41 eV, rendering these nanocomposites conducive for contaminant removal. The nanocomposite demonstrated heightened efficacy in degrading the pollutant erythrosine, achieving a degradation efficiency of 92.9%. While these advantages confirm that ceramic-based nanocomposites are valuable for pollutant degradation, it is essential to consider the context, pollutant characteristics, and treatment goals when selecting nanocomposite materials, acknowledging that other types such as polymer-based or metal-based nanocomposites may also be suitable for certain applications, depending on the specific requirements.

4.2.1.4 Carbon-based nanocomposites

Carbon-based nanocomposites that incorporate nanoparticles such as carbon nanotubes or graphene are designed to efficiently degrade pollutants. Nanomaterials characterized by distinctive nanoscale attributes can be tailored to enhance either catalytic or adsorption functionalities for specific pollutants. In environmental remediation, these nanocomposites interact with pollutants, improving degradation or aiding removal. Leveraging the high surface area and reactivity of carbon-based nanomaterials proves advantageous in wastewater treatment, air purification, and soil remediation [36]. Graphene, along with its derivatives including graphene oxide (GO) and reduced graphene oxide (rGO), constitutes a class of carbon-based materials that act as efficient substrates for the dispersion of nanostructured metal oxides. These platforms play a crucial role in averting electron–hole pair recombination, consequently enhancing photocatalytic activity. GO, acting as a semiconductor photocatalyst, undergoes gradual bandgap shifts during the catalytic process due to the loss of oxygen functionalities, resulting in enhanced conductivity. Recently, Mohamed *et al* [37] utilized hydrothermal synthesis to fabricate nanocomposites comprising Ag–CdSe/GO/CA, revealing heightened photocatalytic activity in the degradation of malachite green under UV–visible light conditions. The enhanced performance of the composite resulted from the multifunctional roles played by GO, silver (Ag), and cellulose acetate (CA) in providing structural support. The GO served as an electron acceptor, promoting carrier separation, while copper nanoparticles on semiconducting rGO exhibited notable degradation of organic dyes [38]. Similarly, Rosalin *et al* [39] synthesized nanocomposites of Ag-loaded ZnO-graphene (AgZG) utilizing hydrothermal techniques, resulting in notable degradation of MO under both UV radiation and sunlight. The heightened photocatalytic performance was ascribed to the judicious loading of Ag, which facilitated the enhanced production of $OH^{\bullet}$ radicals, augmented the absorption of visible light, and increased the efficiency of the charge transfer processes. Gu *et al* [40] synthesized graphene-doped TiO_2 using the hydrothermal method; their results revealed that the graphene played a critical function in sequestering generated electrons, inhibiting electron–hole recombination, enhancing the capacity for pollutant adsorption, and extending the range of light absorption wavelengths. Li *et al* [41] fabricated a nanocomposite photocatalyst comprising silver/bismuth oxyiodide/graphene oxide (Ag/BiOI/GO), which exhibited heightened efficiency in rhodamine B degradation. The study underscored the enhanced photocatalytic capability of the Ag/BiOI/GO, which was attributed to the presence of both Ag and GO in the composite. Fulzele *et al* [42] synthesized a high-performance photocatalyst, an $rGO/Ag_3PO_4/CeO_2$ nanocomposite, using a sonochemical method. The catalyst demonstrated remarkable photocatalytic efficacy in the degradation of MB dye under visible light, achieving close to complete degradation. This exceptional performance was ascribed to the efficient separation of electron–hole pairs, underscoring the nanocomposite's stability and effectiveness.

Carbon nanotube nanocomposites are recognized for their efficacy in pollutant degradation, which is due to their expansive surface area, elevated conductivity, and chemical stability. Some of these nanocomposites exhibit photocatalytic activity and can be customized for tailored solutions, often synergizing with other nanomaterials for enhanced performance. For instance, polyaniline/multiwalled carbon nanotube nanocomposites (PANI/MWCNT NCs) were prepared by Ardani *et al* [43] using the ultrasonic method. The outcomes demonstrated that elevating the ultrasonic power and duration resulted in a reduction in the nanocomposites' particle size, which ranged from around 190–103 nm, accompanied by a specific surface area of 25.34 $m^2\ g^{-1}$. Photodegradation investigations revealed that PANI/MWCNT NCs containing 7 wt% of MWCNTs demonstrated notable efficacy, achieving a removal efficiency of 98.39% for MO dye within 45 min. This surpassed the performance of pure PANI, which achieved a removal efficiency of 79.80%. Kinetic analysis indicated a good fit with the pseudo-second-order model. In another study conducted by Rad *et al* [44], zinc chromium layered double hydroxide (ZnCr-LDH) nanocomposites were synthesized by integrating CNTs and bamboo charcoal (BC), resulting in an increase in specific surface areas, as determined through BET analysis. The initial surface area of the ZnCr-LDH was 1.2 $m^2\ g^{-1}$, which was significantly enhancement to 5.0 and 7.0 $m^2\ g^{-1}$ upon integration with CNTs and BC, respectively. The hybridized materials manifested decreased bandgap values of 2.46, 2.22, and 1.60 eV, indicating enhanced electron–hole pair generation under ultrasonic irradiation and light sources. The hybrid materials exhibited noteworthy catalytic efficacy in the sonophotocatalytic degradation of rifampicin (RF), achieving 100% degradation efficiency within 40 min when exposed to 150 W of ultrasound and irradiation with visible light at a concentration of 0.6 g L^{-1} ZnCr-LDH/BC. The use of CNT composite photocatalysts for wastewater treatment is a promising approach, but it faces various challenges. High production costs, due to the expensive synthesis of quality CNTs, hinder its large-scale implementation in regions with budget constraints. CNTs tend to aggregate together, reducing their surface area and hindering their dispersion in the photocatalyst matrix. Concerns about the toxicity of CNTs in wastewater treatment raise questions about their environmental impact and long-term risks. Photocorrosion, which is influenced by harsh conditions, may decrease the photocatalyst's activity over time. Limited absorption of visible light affects efficiency, and scaling up production faces technical challenges. It is crucial to improve the selectivity for pollutant removal without impacting non-target compounds

Fullerenes, notably C60, are unique carbon allotropes with a spherical or ellipsoidal structure. Discovered in 1985, they blend particle and wave properties, conducting electricity and exhibiting light absorption/emission. Widely used, especially in Fullerene nanocomposite photocatalysts, they enhance light-driven catalysis in applications such as environmental remediation and energy production. A Fullerene nanocomposite photocatalyst comprises fullerene nanoparticles within a semiconductor host matrix, such as titanium dioxide (TiO_2) or zinc oxide (ZnO). Fullerenes such as C60 or C70 which have unique light-absorbing properties enhance the photocatalyst's performance. In these nanocomposites, fullerenes act as sensitizers, absorbing light to generate and efficiently separate electrons. The semiconductor matrix facilitates redox reactions upon photon absorption, which is crucial for pollutant degradation or hydrogen production. A recent study conducted by Zhang *et al* [45] synthesized a hybrid nanocomposite comprising titanium dioxide (TiO_2) and fullerene using an ultrasonication–evaporation technique. The resulting TiO_2/fullerene composite, consisting of well-dispersed nanoparticles around 25 nm in size, demonstrated high efficiency (over 96%) in degrading rhodamine B dye under visible light ($k > 400$ nm). The introduction of fullerene into TiO_2 extended the absorption spectrum of the photocatalyst into the visible light domain, concurrently increasing adsorption capacity and degradation efficiency. The observed improvement was attributed to the synergistic influence arising from the interaction between fullerene and TiO_2, making the composite a promising, efficient, and reusable photocatalyst for environmental applications [45]. Fullerene nanocomposite photocatalysts offer advantages such as tunable bandgaps and efficient charge separation, making them strong contenders for certain applications. However, their costly synthesis, limited solubility, and potential toxicity pose challenges. Comparatively, graphene and carbon nanotubes provide high surface area but struggle with charge separation. Carbon black's high cost and limited surface area restrict its performance. The optimal carbon-based photocatalyst choice depends on application-specific requirements, including factors such as cost, stability, and potential toxicity. In conclusion, carbon-based photocatalysts have a promising future in pollutant degradation. These materials exhibit exceptional catalytic properties, paving the way for sustainable environmental remediation. As research progresses, we foresee ongoing progress in the evolution of carbon-based photocatalysts characterized by enhanced efficiency and cost-effectiveness. This innovation holds the potential to revolutionize pollutant abatement technologies, offering a green and scalable solution that can address environmental challenges.

4.2.2 Methods of nanocomposite synthesis

Nanocomposite photocatalysts represent a cutting-edge advancement in materials science, particularly in the realm of environmental remediation and energy conversion. Two principal methodologies, top-down and bottom-up, are utilized for their preparation. The top-down strategy entails breaking larger materials into nanoscale components, offering control over size and composition but facing challenges in maintaining uniformity [46]. Conversely, the bottom-up method builds nanocomposites from individual molecules or nanoparticles, providing precise control over structure and composition for enhanced efficiency [46, 47]. Combining these approaches holds promise for optimizing nanocomposite photocatalyst design and fabrication, which is applicable across diverse fields such as wastewater treatment and solar energy conversion. These nanocomposites composed of nanoscale elements exhibit superior performance in photocatalysis because of their enhanced surface area, improved charge separation, and synergistic effects. Their preparation involves the integration of various nanomaterials, such as semiconductor nanoparticles and metal oxides, using a number of techniques like sol–gel synthesis, hydrothermal methods, etc. Crucial factors include the selection of the precursors and the optimization of the reaction conditions, which enable photocatalysts to be customized for specific applications such as pollutant degradation and solar energy conversion [48]. Continued research into nanocomposite photocatalyst preparation holds promise for addressing environmental sustainability and clean energy challenges. Various techniques, including precipitation, sol–gel, hydrothermal, and sonochemical methods, are utilized in the synthesis of nanocomposites. Table 4.1 examines the influences of the synthesis method and process conditions on the characteristics of the resulting photocatalyst. The choice of method is contingent upon factors such as the nature of the nanocomposite, the simplicity of the method, and the specific properties of the nanocomposite.

4.2.2.1 The precipitation method

The precipitation method is a commonly used technique for synthesizing nanocomposites. It involves mixing two or more precursor solutions that contain the desired components of the nanocomposite. When these solutions are mixed, a chemical reaction takes place, resulting in the formation of insoluble particles that precipitate out of the solution. These particles then act as building blocks for the nanocomposite. By adjusting factors such as the precursor concentration, temperature, and pH during synthesis, it is possible to regulate the dimensions, composition, and structural attributes of the produced nanoparticles. After precipitation, additional processing steps may be necessary to improve the nanocomposite's properties. These can include washing and drying to remove impurities and solvents. The procedure used for synthesis is depicted in figure 4.3, which shows a detailed

Table 4.1. The impacts of the preparation method and the process conditions on photocatalyst properties.

Photocatalyst nanocomposite	Preparation method	Reaction conditions	Properties of the nanocomposite	Results	Ref.
MgO/GO	Hydrothermal method	• 1:1 ratio mol % of GO and MgO • 1 h sonication • 4 h placed in autoclave at 80 °C	• Structure: flaky spherical • Nanoparticle size: 12 nm (average) • Bandgap: 1.7 eV	• MgO/GO photocatalyst demonstrated exceptional photocatalytic activity (98%). • Sustained high performance consistently observed over five consecutive cycles.	[61]
$AgIO_4/CeO_2$	Sol–gel and sonochemical route	• Ultrasonic bath: 200 W power, 1 h duration • Wet solid subjected to ultrasonic treatment • Dried in oven: 12 h	• Specific surface: 49 $m^2\ g^{-1}$ • Crystalline size: 17 nm • Bandgap energy: 1.98 eV	• RhB dye decomposition: 95% complete • Absence of secondary pollutant generation • Promising results for sustainable wastewater treatment	[62]
CuO/NiO	Coprecipitation	• Mixture pH: 10 • Dried product calcination: 400 °C, 5 h	• Nanocomposite-specific surface area measured 22.28 $m^2\ g^{-1}$. • Pore size: 3 nm • Bandgap: 4.68 eV	• Cefixime degradation: 90% • Kinetics: first-order • Rate constant (k): $7.4 \times 10^{-3}\ min^{-1}$	[63]
CuO/ZrO_2	Sol–gel method	• Reaction temperature: 40 °C • Calcination temperature: 500 °C for 4 h	• Surface area measured 200 $m^2\ g^{-1}$, • Bandgap was determined to be 2.72 eV. • Average particle size: 45 nm	• Photocatalytic efficiency for thiophene oxidation: 100% in 120 min • Rate constant of CuO/ZrO_2 nanocomposite: 0.0172 min^{-1} • Enhancement compared to ZrO_2 NPs: 3.5-fold increase	[64]

rGO/TiO_2	Hydrothermal method	rGO/TiO_2 precursor solution ultrasonically treated: • Power: 600 W • Room temperature • Duration: 1 h • Hydrothermal treatment: • Temperature: 1200 °C • Duration: 18 h	• Crystalline size: 15 nm • Bandgap: 2.02 eV	• Degradation rate: 13.54%	[65]
CuO/TiO_2	Ultrasound-assisted method	• Ultrasonic wave frequency: 20 kHz • Room temperature	• Shape: spherical • Particle size: 20 nm • Bandgap: 3.08 eV	• Highest decay constant observed: 1.38 min^{-1} • Target: sulfate-reducing bacteria • Significant enhancement in degradation efficiency	[66]
ZnO/CuO/Ag	Coprecipitation	• pH maintained at 9 • Precipitate annealed at 500 °C	• Nanocomposite photocatalyst bandgap: 2.85 eV • Spherical nanoparticles; average size: 140 nm • Aggregation of nanoparticles observed • Specific surface area: 21.3 $m^2\ g^{-1}$	• The ZnO-CuO-Ag nanocomposite demonstrated improved efficiency in charge separation. • Achieved 100% photodegradation of MB after 90 min of visible light radiation. • Photocatalytic reaction rate constant measured at 0.0515 min^{-1}.	[67]
TiO_2/GO	Modified sol–gel method	• Reaction time: 1 h • Room temperature • Calcination temperature: 500 °C • Calcination duration: 5 h • Heating rate during calcination: 20 °C min^{-1}	• Specific surface area: 46.71 $m^2\ g^{-1}$ • Bandgap shift: 3.07 eV to 2.92 eV due to GO incorporation • Nanocomposite size: 30 nm	• MB dye degradation: up to 98% • pH ≈ 13 • Optimal degradation time: 60 min	[68]

Table 4.1. (*Continued*)

Photocatalyst nanocomposite	Preparation method	Reaction conditions	Properties of the nanocomposite	Results	Ref.
ZnO/SnO_2/rGO	Hydrothermal method	• GO synthesized using Hummer method • Nanocomposite precursor mixture sonicated for 30 min • Solution heated to 130 °C for 18 h	• Bandgap energies: ZnO: 2.8 eV; ZnO/rGO: 2.3 eV; SnO_2/rGO: 2.0 eV; ZnO/SnO_2/rGO: 1.9 eV • Average particle size of ZnO/SnO_2/rGO nanocomposites: 180 nm	• Orange dye degradation rate constant: 0.0241 min^{-1} • Nanocomposite: ZnO/SnO_2/rGO • Degradation efficiency: >91%	[69]
NiO/RGO	Hydrothermal method	• Teflon stainless autoclave used • Incubation at 180 °C for 10 h • Formation of black NiO/RGO nanocomposite powder • Annealed for 6 h • Annealing temperature: 350 °C	• Nanocomposite bandgap: 3.47 eV • Average crystal size: 22.5 nm.	• Photocatalytic degradation efficiency: 95.4% • Nanocomposite: NiO/RGO • Rate constant: 0.015 871 min^{-1} • NiO/rGO rate constant 1.4 times higher than that of NiO	[70]
ZnO/MgO/CaO	Biosynthesis	Initial solution: • Temperature: 80 °C. Duration: 8 h Homogeneous solution: • Calcination temperature: 700 °C • Calcination duration: 2 h	• Nanocomposite exhibits crystalline nature • Triangular shapes with a size of approximately 59 nm • Nanoparticle optical bandgap measures 5.3 eV	• MB dye degradation: 93% • Duration of UV light exposure: 120 min • Rate constant (kobs): 0.007 min^{-1}	[71]

$ZnO/TiO_2/rGO$	Laser ablation method	Laser parameters: • Pulse energy: 130 mJ • Wavelength: 355 nm • Repetition rate: 10 Hz • Pulse width: 10 nsLens type: UV lens, focal length: 200 mm • Irradiation time: 30 min	• Nanocomposites exhibit a low bandgap of 2.73 eV. • 355 nm laser irradiation transforms spherical ZnO nanoparticles into nanorods. • Morphology becomes more compact post-irradiation.	• Rate constant: 0.149 min^{-1} • $ZnO/TiO_2/rGO$ nanocomposite achieved >98.5% degradation of MB within 60 min • Flat sheet surface area increased with higher rGO content • The presence of ZnO and TiO_2 contributed to enhanced surface area	[72]

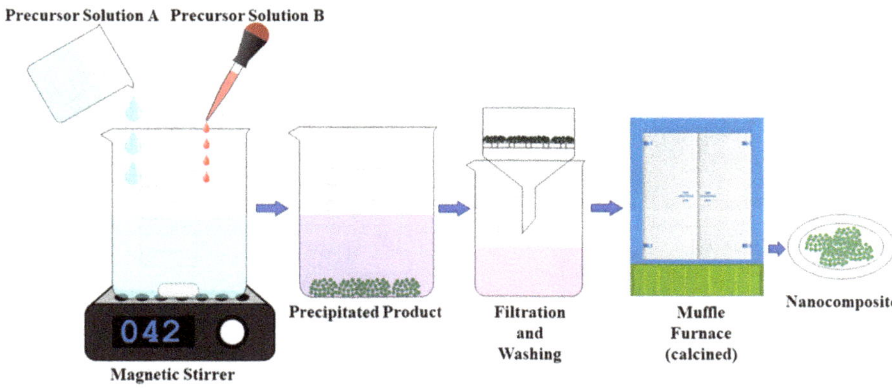

Figure 4.3. A step-by-step schematic representation of the precipitation method.

schematic representation of the precipitation method. Researchers can also treat the nanoparticles or combine them with other materials to create a composite with specific characteristics. The precipitation method is widely used because it is simple and cost-effective and gives control over particle size and composition. It is scalable, making it suitable for industrial-scale production. The homogeneous distribution of nanoparticles within the composite ensures uniform photocatalytic activity.

In a study, Wu *et al* [49] effectively produced a nanocomposite of Ag_2CO_3/ZnO using a chemical precipitation method. Zinc oxide nanowires weighing 1 g and corresponding to 12.3 millimoles were dispersed in 500 milliliters of distilled water. Following the introduction of silver nitrate ($AgNO_3$ 0.067 g, 0.39 mmol) in NH_4OH (2.3 ml, 25 wt% NH_3), the nanowire suspension was sonicated for 30 min. The resulting suspension was magnetically stirred for 30 min at room temperature, followed by the rapid addition of sodium bicarbonate (0.5 mmol) in 50 ml of deionized water, which was continuously stirred. Following 12 h of room-temperature stirring, the precipitates were filtered, sequentially washed with distilled water and ethanol, and then dried at 80 °C for 12 h. The resulting Ag_2CO_3/ZnO nanocomposite exhibited a surface area of 73.2 $m^2\ g^{-1}$ and demonstrated significant efficacy, achieving 91% degradation of rhodamine 6G (R6G) under visible light exposure in 40 min. A study carried out by Haghighatzadeh *et al* [50] employed a straightforward chemical coprecipitation technique to synthesize ZnO/TiO_2 nanocomposite catalysts featuring adjustable TiO_2 shell thickness. The resulting photocatalysts exhibited distinct crystal structures, namely wurtzite ZnO and anatase TiO_2 as the core and shell, respectively. Their bandgap energy was 3.05 eV following the incorporation of TiO_2, a decrease from the original value of 3.4 eV. The results revealed the surface morphologies of both ZnO nanoparticles (NPs) and the ZnO/TiO_2 nanocomposites. A field emission scanning electron microscopy (FESEM) image of the ZnO NPs displayed nonuniform NPs with sheet-like shapes attributed to agglomerated spheres. The particle sizes, as illustrated in figure 4.4, ranged from 18 to 50 nm. Conversely, the ZnO/TiO_2 core–shell structures exhibited relatively

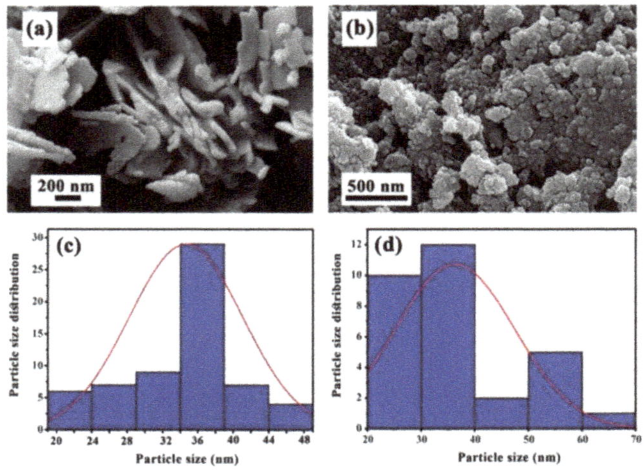

Figure 4.4. Images (a) and (b): field emission scanning electron microscopy (FESEM) visualizations. (c) and (d): the particle size distributions of ZnO and the ZnO/TiO_2 nanocomposites, respectively. Reproduced from [50].

uniform spherical particles with an average dimension of roughly 30 nm, indicating the effectiveness of the precipitation method in achieving homogeneous structures.

In another study, Haounati *et al* [51] successfully prepared a Fe_3O_4/Ag_3PO_4@Sep ('Sep' denotes sepiolite) magnetic nanocomposite through an eco-friendly one-step precipitation method. This nanocomposite exhibited increased photocatalytic efficacy in the degradation of malachite green (MG) dye under sunlight conditions; the most active composite completely removed MG dye within 16 min at a 10 ppm concentration. The nanocomposite exhibited degradation times that were three, five, and six times faster than those of Ag_3PO_4@Sep, Ag_3PO_4, and Fe_3O_4, respectively. The hybrid configuration exhibited a decreased bandgap energy of 2.37 eV, which improved its photocatalytic capability under visible light. Significantly, the continuous photocatalytic activity of Fe_3O_4/Ag_3PO_4@Sep continued even following three cycles of reuse. Precursor selection allowed nanoparticles to be synthesized with specific properties, enhancing the photocatalyst's effectiveness. This method is adaptable to diverse substrates, including polymers and ceramics, which provides flexibility for various applications. Factors such as ease of synthesis and scalability should be carefully considered when choosing a method for nanocomposite photocatalyst synthesis.

4.2.2.2 The hydrothermal method

The hydrothermal technique is a crucial method for synthesizing nanocomposite photocatalysts; it involves synthesis at elevated pressure and temperature in an aqueous environment. Starting with a precursor solution containing the desired nanomaterials, such as semiconductor nanoparticles or metal oxides, contents of a sealed reaction vessel undergo controlled growth and assembly under extreme conditions. This method facilitates precise control over size, morphology, and

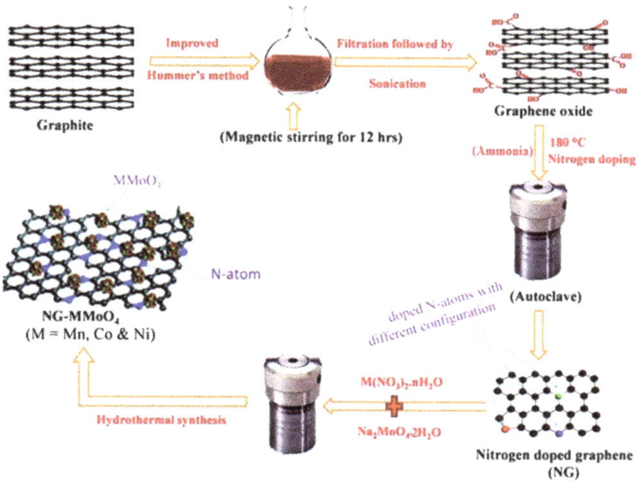

Figure 4.5. A representation of the synthesis of graphene oxide, nitrogen-doped graphene, and the nanocomposite NG-$MMoO_4$. Reprinted from [52], Copyright (2017), with permission from Springer Nature.

composition, resulting in well-defined structures with enhanced photocatalytic activity. In a recent investigation conducted by Kumar *et al* [52], the hydrothermal approach was employed for the synthesis of nanocomposites consisting of nitrogen-doped graphene combined with metal molybdates (NG-$MMoO_4$), in which $MMoO_4$ represents a metal molybdate and M denotes Mn, Co, or Ni. In this process, nitrogen-doped graphene was sonicated and centrifuged to obtain a colloidal solution. Sequential additions of $M(NO_3)_2{\cdot}nH_2O$ (where M = Mn, Co, Ni) and $Na_2MoO_4{\cdot}2H_2O$ were made and the pH was adjusted to 7. The resultant suspension was transferred to a Teflon-lined stainless steel autoclave and subjected to heating at 180 °C for a period of 12 h, followed by subsequent cooling. The synthesis process is depicted in figure 4.5, which schematically represents each step in the procedure. The nanocomposites displayed nanostructures characterized by crystallite sizes ranging from 10 to 30 nm and demonstrated a remarkable 99% degradation of MB.

Similarly, another work carried out by Younas *et al* [53] successfully synthesized a $WO_3/TiO_2/V_2O_5$ catalyst using a hydrothermal method. Precursor solutions were combined in a stainless steel Teflon-lined autoclave and maintained at a temperature of 120 °C for 24 h. Subsequently, the resultant precipitates were washed, filtered, centrifuged, and then calcinated at 620 °C for 4 h within a muffle furnace to produce the nanocomposite. The results revealed varied structural morphologies with dimensions varying between 50 and 80 nm, and the nanocomposite displayed an average particle size of 20.44 nm. It demonstrated notable photocatalytic efficacy, yielding degradation percentages of 76%, 86%, and 86% for MB, MO, and orthonitrophenol, respectively, highlighting its excellent potential. Similarly, a study carried out by Parasuraman *et al* [54] synthesized $FeWO_4/FeS_2$ nanocomposites via the hydrothermal method and revealed that the 2:1 wt% $FeWO_4/FeS_2$ nanohybrid established a heterojunction characterized by the lowest recombination rate of electron–hole pairs and minimal electron transfer resistance. A Tauc plot analysis determined a bandgap value of 2.76 eV for $FeS_2/FeWO_4$. The nanohybrid demonstrated superior degradation activity, attaining an efficiency of 89.5% in 120 min under UV–visible irradiation. This outcome was ascribed to synergistic effects, enhanced light absorption, and proficient separation of charge carriers. Also, a novel hybrid nanocomposite comprising multiwalled carbon nanotubes (MWCNTs) with copper oxide, as well as pristine copper oxide, were synthesized via hydrothermal techniques by Shibu *et al* [55]. The MWCNTs (50 wt%) were dispersed in a solution of copper nitrate at a concentration of 0.5 M and a volume of 100 ml under continuous stirring. The pH was adjusted to 13 by adding 0.2 M NaOH. The resultant blend was subjected to hydrothermal processing at 120 °C for 15 h, subsequently undergoing centrifugation, rinsing, and heating at 150 °C for 15 h to yield the final MWCNT/copper oxide nanocomposite. A comparative investigation into the inclusion of MWCNTs versus pristine copper oxide unveiled noteworthy alterations in the crystalline phase, lattice parameters, particle characteristics, morphology, and the optical bandgap. The nanocomposite of MWCNTs and copper oxide exhibited a measured bandgap of 1.90 eV. This hybrid material displayed heightened efficacy in the photodegradation of methyl orange (MO) dye when exposed to sunlight for 6 h. The degradation process followed pseudo first-order kinetics, resulting in the breakdown of 92% of the MO dye. The MWCNT/copper oxide nanocomposite maintained a high level of recyclability, demonstrating consistent performance for up to five cycles. With the hydrothermal method, widely applied in materials science, the future of hydrothermal synthesis for the preparation of nanocomposite photocatalysts holds exciting prospects. The hydrothermal method is advantageous for synthesizing nanocomposite photocatalysts, but faces some challenges. Control of the particle size, shape, and morphology is difficult due to factors such as temperature and pressure. High temperatures and pressures may introduce impurities, demanding extra purification steps. Energy consumption is a concern for large-scale production. Some precursors are incompatible, limiting nanocomposite diversity. Certain solvents raise environmental concerns. Despite these challenges, ongoing research focuses on optimizing hydrothermal synthesis, choosing suitable precursors, and developing eco-friendly alternatives. Combined techniques, the use of computational modeling, and innovative strategies are being explored to enhance control and efficiency in nanocomposite photocatalyst synthesis. Researchers can expect to delve deeper into optimizing conditions, designing tailored nanocomposites, scaling up production, integrating with emerging technologies, and exploring new nanomaterials to unlock the full potential of these materials for sustainable and efficient applications.

4.2.2.3 The sonochemical method

The sonochemical method for preparing nanocomposite photocatalysts includes utilizing high-frequency ultrasound waves to induce acoustic cavitation in a reaction mixture. This process generates localized intense heating, high pressures, and shockwaves, promoting the formation of nanoparticles from larger precursors. Ultrasound exposure induces the collapse of bubbles, resulting in the formation of

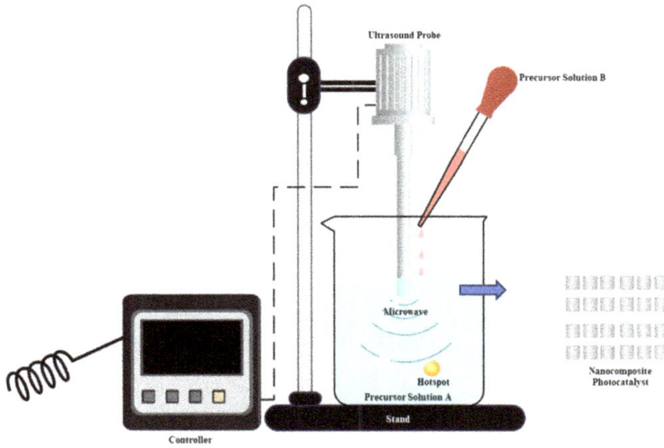

Figure 4.6. Presents a schematic illustration delineating the application of the sonochemical method in the synthesis of nanocomposites.

localized regions characterized by intense heat (5000 K) and elevated pressure (20 MPa). This non-hazardous and rapid method produces small metal particles efficiently. Figure 4.6 shows a schematic representation of the setup used in the sonochemical method for synthesizing nanocomposites. Using high-frequency waves, it is possible to precisely control the size, morphology, and composition. This approach, conducted under mild conditions, holds promise for environmentally friendly synthesis, enhancing nanocomposite properties for applications in environmental remediation and energy conversion.

In a recent study, Zinatloo-Ajabshir *et al* [56] successfully prepared a nanocomposite of $ZnCo_2O_4/Co_3O_4$ using a sonochemical process. This composite showed great potential for efficiently removing organic contaminants. According to the study, the optimal conditions with ultrasound for producing the nanocomposite with the smallest and most uniform particles were a synthesis time of 10 min and the use of cetyltrimethylammonium bromide (CTAB) as a capping agent. The outcomes revealed that the systematically optimized nanocomposite, formulated with CTAB over a duration of 10 min, exhibited notable removal efficiencies for three distinct contaminants, namely methyl violet (100%), Acid Red 14 (95.6%), and erythrosine (100%). This was achieved within just 65 min under visible radiation. The optimized sample was also found to be durable and could maintain its catalytic yield for up to eleven runs without significant loss of effectiveness. A nanocomposite of Ag_2O/CuO was synthesized via ultrasonication by Warsi *et al* [57] exhibiting crystallite sizes of 17.55 nm for Ag_2O and 17.42 nm for CuO. Tauc plots were utilized to ascertain a diminished bandgap energy of 1.51 eV for the nanocomposite. This reduction signifies favorable alterations in the electronic structure, enhancing its photocatalytic capabilities under solar irradiation when contrasted with pristine CuO (with a bandgap of 2.77 eV) and Ag_2O (with a bandgap of 1.61 eV). The nanocomposite demonstrated superior photocatalytic activity, with percentage degradation values of 97.5%, 90.5%, and 95% for MB, RB, and benzoic acid, respectively, surpassing pure metal oxides.

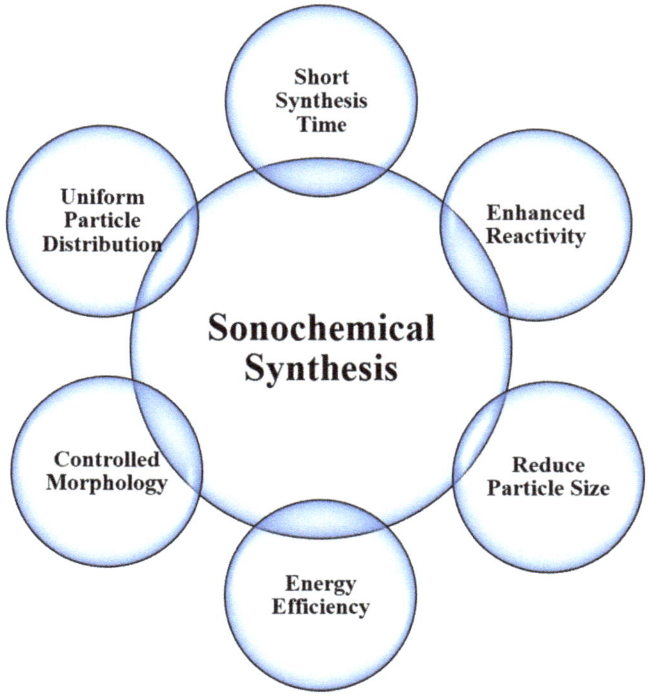

Figure 4.7. Advantages of sonochemical synthesis of nanocomposite

Sonochemical synthesis of nanocomposite photocatalysts presents several advantages shown figure 4.7, including uniform nanoparticle distribution through intense cavitation-induced mixing. This method enhances reactivity by generating high-energy conditions, fostering the formation of reactive species for improved nanocomposite structures. Sonochemistry reduces particle size, increasing surface area for better photocatalytic performance, and requires shorter synthesis times. Precise control over morphology is achieved by adjusting synthesis parameters, ensuring tailored properties for specific applications. The process is energy-efficient, operating at ambient temperatures, and exhibits versatility with various precursor materials. Sonochemical synthesis is scalable, facilitating the production of nanocomposite photocatalysts in larger quantities for practical use.

4.2.2.4 The sol–gel method

The sol–gel methodology is extensively utilized and exhibits versatility as a technique for the synthesis of nanocomposites. It involves the dissolution of metal salts or metal alkoxides in a solvent, leading to the formation of a colloidal suspension (sol), which then undergoes controlled hydrolysis and condensation reactions to transition into a solid gel phase. This process occurs at relatively low temperatures and allows for the precise dispersion of nanoscale components such as nanoparticles or nanotubes within a solid matrix. The resulting nanocomposite,

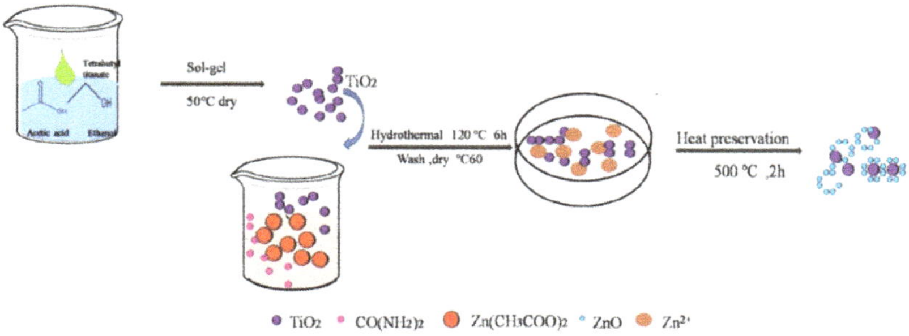

Figure 4.8. Representation of the synthesis of the TiO_2/ZnO nanocomposite through the sol–gel method. Reprinted from [59], Copyright (2021), with permission from Springer Nature.

often in gel form, offers researchers the ability to tailor its composition and structure for specific applications, such as photocatalysis. By manipulating parameters such as the precursor choice, solvent, and reaction conditions, precise control over the nanocomposite's properties can be achieved. This method's utility in creating functional materials, such as those used for photocatalysis, is evident. In an investigation conducted by Pudukudy *et al* [58], the sol–gel methodology was applied to fabricate ultrafine nanospherical composites of CeO_2/TiO_2. The introduction of 1% CeO_2 augmented the specific surface area and enhanced visible-light absorption, thereby diminishing the optical bandgap of TiO_2. The heterojunctions formed in the 4% CeO_2/TiO_2 nanocomposite substantially mitigated electron–hole pair recombination, thereby increasing photocatalytic efficiency with a tetracycline (TC) removal efficiency of ~99% after 80 min of sunlight exposure. The enhanced performance was attributed to reduced particle size, increased particle surface area, heightened absorption of visible light, reduced bandgap, and prolongation of the lifespan of photogenerated charge carriers facilitated by the formation of a heterointerface. Likewise, in a study by Bai *et al* [59], a TiO_2/ZnO nanocomposite was synthesized utilizing sol–gel methodologies. The procedural approach involved the synthesis of TiO_2/ZnO nanocomposites, as shown in figure 4.8. They diluted $Cl_6H_{36}O_4Ti$ in ethanol and added deionized water and acetic acid sequentially with constant agitation to generate a sol, which, when heated to 50 °C, transformed into a gel. The precursor was dried at 80 °C and calcinated at 550 °C for 3 h to obtain TiO_2. CH_4N_2O, Zn $(CH_3COO)_2{\cdot}2H_2O$, and the prepared TiO_2 were mixed, dissolved in deionized water, heated to 120 °C for 6 h, centrifuged, and finally calcinated at 500 °C for 2 h to yield TiO_2/ZnO. The nanocomposite displayed a decreased bandgap width of 2.89 eV and featured a surface area of 15.28 $m^2\ g^{-1}$. Its photocatalytic degradation rate of MB was remarkable, reaching ~100% at 80 min; the first-order kinetic equation was followed, characterized by a kinetic constant of 0.037 48 min^{-1}. Notably, this the magnitude of this constant was 2.8 times higher than that of pure ZnO and 4.8 times higher than that of pure TiO_2.

Akhter *et al* [60] conducted an investigation involving the synthesis of a ZnO/TiO_2 nanocomposite through the application of the sol–gel method. The

Fig. 1. SEM images (a), (b), (c) TiO_2 nanoplates (d), (e), (f) ZnO nanoparticles (g), (h), and (i) ZnO/TiO_2 nanocomposites.

Figure 4.9. Scanning electron microscopy (SEM) images (a), (b), and (c) show TiO_2 nanoplates, while (d), (e), and (f) represent ZnO nanoparticles. Images (g), (h), and (i) illustrate the ZnO/TiO_2 composite. Reprinted from [60], Copyright (2023), with permission from Elsevier.

methodology involved the mixture of 3 g of ZnO with 10 ml of ethanol and 5 ml of distilled water, which were stirred for 15 min. Simultaneously, a mixture of 3 g of TiO_2, 10 ml of ethanol, and 5 ml of distilled water was stirred for 10 min. Five ml of concentrated HCl was mixed into this TiO_2 solution, stirred for 15 min, and gradually introduced into the ZnO solution while maintaining a pH of more than 1.5. The resultant mixture was stirred for 3 h at ambient temperature, followed by exposure to a hot water bath at 100 °C for 1 h and subsequent cooling. After 30 min of centrifugation, the nanoparticles were filtered, oven dried, dehydrated, and then subjected to calcination in a muffle furnace at 400 °C for 3 h. The synthesized ZnO/TiO_2 nanocomposite exhibited a spherical morphology characterized by diameters ranging from 300 to 500 nm. The scanning electron microscope (SEM) images depicted in figure 4.9 elucidate the surface morphologies of the synthesized nanoparticles and nanocomposites. The photocatalytic performance of ZnO/TiO_2 surpassed that of the pristine photocatalysts, resulting in a 98% degradation of MB dye, and a kinetic analysis of the photocatalysts indicated the presence of pseudo first-order kinetics.

The sol–gel methodology is commonly employed in the preparation of nanocomposite photocatalysts intended for the treatment of wastewater; nevertheless, it encounters numerous challenges. The process is complex and requires critical control of the pH, temperature, and precursor concentrations. This can influence the uniformity of the ultimate product. Achieving nanoparticle homogeneity is also difficult; lack of homogeneity results in uneven distribution and reduced photocatalytic activity. The material's pore structure and surface area may not conform to ideal specifications, which can impact reactant accessibility. The drying and gelation process can cause cracking, reducing the structural integrity of the material. Additionally, scaling up for industrial production is a challenge, and high-temperature processing is energy-intensive and thus raises concerns. Chemical residues and the high costs attributed to expensive precursors are potential drawbacks. Material compatibility issues further limit the method's applicability, despite its advantages in tailoring material properties. Ongoing research is being conducted to address these limitations.

In conclusion, the synthesis of nanocomposite photocatalysts is a dynamic and evolving field that is crucial for advancing sustainable technologies. Scientists utilize various methodologies, including sol–gel, precipitation, hydrothermal, and sonochemical processes, each of which offers unique advantages. The sol–gel method provides precise control over composition, precipitation methods offer scalability, hydrothermal synthesis produces crystalline structures, and sonochemical methods enable rapid synthesis. These techniques, when combined, result in synergistic effects, enhancing the catalytic activity, stability, and selectivity of materials. Despite the progress made to date, ongoing research is essential to refine methods, address challenges, and explore novel possibilities. Collaborative efforts within the scientific community will drive the field forward, deepening our understanding of nanocomposite photocatalyst synthesis. Realizing the complete capabilities of this method will extend its applications in environmental remediation, energy conversion, and other domains, ensuring a more sustainable and promising future.

4.2.3 Properties of nanocomposite photocatalysts

Nanocomposite photocatalysts exhibit remarkable properties across multiple domains. Mechanically, their nanoscale structure provides increased surface area, enhancing mechanical strength and durability. Electrically, these nanocomposites often demonstrate improved conductivity, facilitating efficient charge transfer during photocatalytic reactions. Optically, the incorporation of nanoparticles enhances light absorption, leading to enhanced photocatalytic activity. Thermally, the nanocomposite structure promotes better heat dissipation, preventing thermal degradation and ensuring sustained performance. In summary, photocatalyst nanocomposites combine enhanced mechanical strength, improved electrical conductivity, superior optical absorption, and efficient thermal management, making them highly effective in various applications, particularly in harnessing light for catalytic processes.

4.2.3.1 Mechanical properties

Nanocomposite photocatalysts exhibit enhanced mechanical properties, primarily attributed to the strong interfaces formed between components, which impede crack propagation. The nanoscale size increases surface area, facilitating improved load transfer and adhesion. Synergistic effects obtained by combining materials with distinct mechanical properties enhance the overall performance [73]. Surface morphology, which influences mechanical interlocking and energy dissipation, plays a pivotal role. The material properties of individual components significantly impact nanocomposites; examples include TiO_2-SiO_2 [74] with enhanced tensile strength due to strong interfaces and CNT–polymer [75] nanocomposites exhibiting heightened tensile strength and wear resistance. Graphene-based nanocomposites showcase exceptional mechanical attributes, due to graphene's unique 2D structure and strong covalent bonds. Applications encompass filtration membranes, structural materials, protective coatings, reinforcing agents, and energy-harvesting devices, exploiting the nanocomposites' strength, durability, and flexibility [76]. The evolving research in this field holds substantial promise, anticipating advancements. Advancements can be made by enhancing our comprehension of the underlying mechanisms and refining specialized synthesis techniques. This research stands to revolutionize industries by offering materials with unprecedented combinations of photocatalytic activity and mechanical strength.

Research into nanocomposite photocatalysts' mechanical properties is rapidly advancing. Understanding enhancements and innovating methods of synthesis can yield tailored performance, revolutionizing industries with unprecedented photocatalytic activity and mechanical strength combinations.

4.2.3.2 Electrical properties

Nanocomposite photocatalysts with unique electrical properties play a pivotal role in pollutant degradation. Key aspects include enhanced conductivity, facilitated charge separation, and transfer through components such as metal nanoparticles, leading to minimized recombination and improved efficiency [76]. Heterojunctions within the nanocomposite generate internal electric fields, directing photogenerated electrons and holes toward the surface, enhancing pollutant degradation [77]. Plasmonic effects introduced by noble-metal nanoparticles enhance light absorption and electron excitation, improving photocatalytic activity. Tuning the nanocomposite bandgap enables tailored photocatalytic activity for specific light wavelengths. Examples include TiO_2-graphene nanocomposites with improved conductivity and Ag-TiO_2 nanocomposites that display enhanced photocatalytic activity [78].

Ongoing research is focused on developing nanocomposites with optimized electrical properties for specific applications, exploring new materials and synthesis techniques, and integrating nanocomposites with other technologies for advanced environmental remediation and sustainable energy production.

4.2.3.3 Thermal properties

The effectiveness and durability of nanocomposite photocatalysts for pollutant degradation hinge on their thermal properties, influencing aspects such as light absorption and charge separation. Notably, high thermal conductivity prevents overheating during the photocatalytic process, ensuring sustained activity. Thermal stability is vital for longevity in high-temperature environments, while low specific heat capacity enables rapid heating and cooling [79]. The nanocomposites' surface area aids efficient heat transfer, averting localized overheating. These characteristics contribute to heightened light absorption, enhanced charge separation, elevated reaction rates, and improved recyclability. Examples include TiO_2-based nanocomposites for wastewater treatment, carbon-based nanocomposites for high-intensity light applications, and Bi_2O_3-based nanocomposites for improved activity at higher temperatures. Challenges include balancing activity with stability and minimizing the environmental impact of high-temperature processes. Overall, manipulating thermal properties is pivotal for developing efficient, durable, and environmentally friendly photocatalysts to combat pollution.

4.2.3.4 Optical properties

The optical properties of nanocomposite photocatalysts play a crucial role in determining their ability to degrade pollutants. These properties govern the nanocomposite's interaction with light and its capacity to produce pairs of electrons and holes, which is essential for the degradation process. The bandgap, which denotes the energy difference between the valence and conduction bands, plays a pivotal role in influencing light absorption, particularly within the visible spectrum, in turn enhancing pollutant degradation. The absorption spectrum, a measure of light absorption at various wavelengths, ensures efficient competition for incident light between the photocatalyst and pollutant [80]. Light scattering within the nanocomposite, while extending the light path length for increased interaction, must remain balanced to maintain overall efficiency. Surface plasmon resonance (SPR) in metal nanoparticles enhances light absorption, benefiting pollutant degradation by elevating electron–hole pair generation. Photoluminescence can be used to assess recombination rates; lower intensity signifies decreased recombination, which is advantageous for pollutant degradation [81]. The tailoring of these optical properties is exemplified in TiO_2-based nanocomposites, where doping narrows the bandgap for visible light absorption, and in metal–semiconductor nanocomposites, where it leverages SPR effects. In addition, graphene-based nanocomposites that have superior conductivity and extensive surface areas enhance light absorption and charge separation, contributing to improved pollutant degradation. This understanding of and control over optical properties offer promising avenues for developing efficient technologies in environmental remediation and sustainable development.

4.3 The fundamentals of photocatalysis

4.3.1 Photocatalysis

The enormous use of chemicals and their entry into bodies of air, soil, and water is leading to an alarming situation. Chemicals from different categories such as dyes, pharmaceuticals, insecticides, pesticides, and volatile organic compounds threaten the environment, even in trace amounts. Emerging contaminants, persistent organic pollutants, and endocrine-disrupting chemicals all have one thing in common, namely their potential to harm life. Across the globe, the problems associated with pollutant degradation are serious. Hence, in light of these considerations, efforts are being made to formulate effective methodologies for the degradation of these pollutants. Photocatalysis is an advanced oxidation technique that successfully degrades pollutants of organic origin. Advanced oxidation methods work by generating hydroxyl radicals and superoxide anions. Amongst the various advanced oxidation methods, such as the Fenton method, the photo-Fenton method, UV photolysis, ozonation, photocatalysis, and sonocatalysis, photocatalysis is widely accepted due to its simplicity and efficiency of operation [82].

Sunlight is a fundamental form of energy responsible for supporting many reactions which are essential for the existence of life on earth [83]. Harvesting the Sun's energy for reactions is the basis of visible-light photocatalysis. UV photocatalysis and near-infrared (NIR) photocatalysis have also been demonstrated by many researchers. Typically, photocatalysis is a photon-induced process with a wide range of applications in the fields of water treatment, light harvesting, hydrogen generation, batteries, self-cleaning materials, soil remediation, sensors, and the synthesis of novel materials. The treatment of waterborne organic pollutants using visible-light active photocatalysts has received considerable focus due to the efficacy and environmental sustainability of photocatalytic operations. The other photocatalysis operations that use UV- or NIR-activated materials have also been found to be efficient in their respective applications [84]. In a catalytic process, a material showing selective response for a particular reaction is chosen as a catalyst, and thus the process is called a catalytic process. Many industrial processes are catalyst based [85]. Photocatalysis is similar to these processes in terms of the selective action of the photoactive material in carrying out the desired process. As a competent approach for treating pollutants, photocatalysis was first applied few decades ago, and using the new nanomaterials created by various novel approaches, the field has expanded ever since [86].

Today, photocatalysis is applied for the treatment of many emerging organic contaminants and persistent pollutants [87]. Other methods involving light such as photolysis [88], the photo-Fenton method [89], photochemical oxidation [90] and photoelectrocatalysis [91] are also applied for the treatment of contaminants. Photocatalysis and photolysis are often studied in parallel in experiments. It is crucial to highlight that photocatalysis involves the utilization of a light-activated material, whereas photolysis simply involves a reaction that takes place in the presence of light. Techniques such as photoelectrocatalysis use an electrochemical cell, whereas photocatalysis does not require an electrochemical cell. Photo-Fenton oxidation involves the oxidation of organic material by Fenton's reagent. In this method, sludge formation is observed. In photochemical oxidation, an oxidant is required along with a light source. Compared to these light-based methods, there are many benefits of photocatalysis, such as the optimum use of chemicals, efficiency of degradation, the ability to harvest light energy, an absence of secondary pollution, and the inclusion of novel materials. Figure 4.10 is a schematic of photon-based methods applied for the degradation of pollutants.

In a photocatalytic process, a photon initiates an interaction with the photocatalyst, inducing electron migration from the conduction band to the valence band. This electron migration is followed by the creation of a hole. Thus, an electron–hole pair is created during photocatalysis. These electrons and holes lead to the formation of species that can degrade pollutants. Although the photon triggers the photocatalytic mechanism, the photocatalyst produces the desired outcome, which depends on the application of the photocatalyst [92].

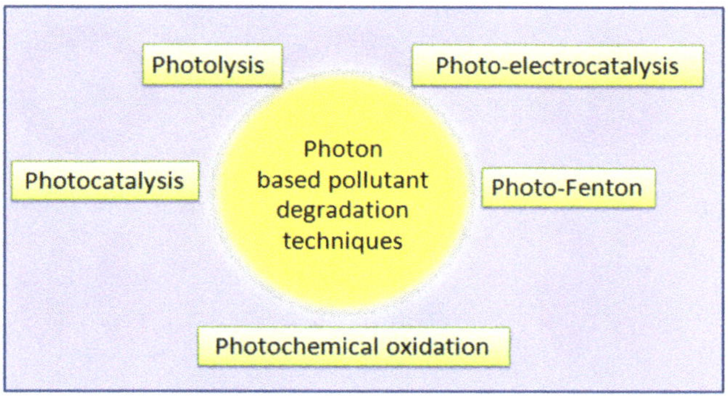

Figure 4.10. Photon-based pollutant degradation techniques

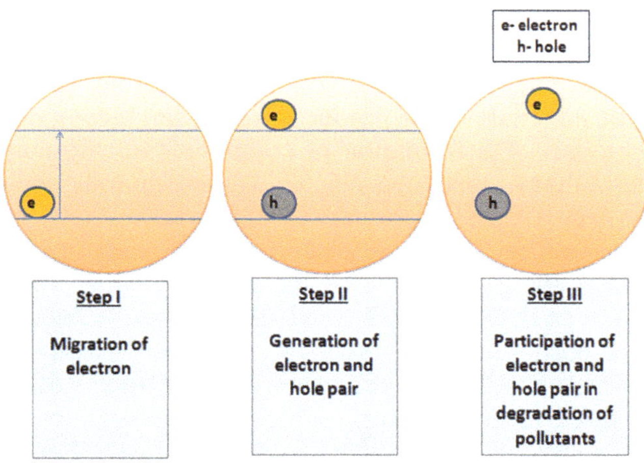

Figure 4.11. A schematic outlining the key stages in the photocatalytic degradation of pollutants.

4.3.2 What are photocatalysts?

Photocatalysts are essentially materials that are able to carry out photocatalysis. This means they are capable of producing electron–hole pairs upon exposure to photons. The three primary stages of photocatalysis are the migration of electrons from the valence band to the conduction band, the generation of electron–hole pairs, and the subsequent participation of electron–hole pairs in pollutant degradation. These steps are shown in figure 4.11.

The research literature describes different types of photocatalysts that are being applied for pollutant degradation. Thus, depending on the light source, photocatalysts may be active under UV, visible, or NIR light [93]. Photocatalysts are based on semiconductor or metal nanomaterials or a combination of these. Figure 4.12 shows the different forms of metal–semiconductor nanocomposites, such as conventional, core–shell, yolk–shell, Janus type, array type, and multi-junction structures [94].

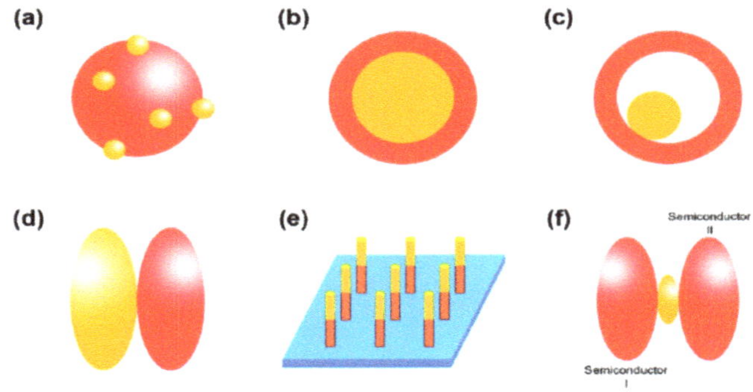

Figure 4.12. The different forms of metal–semiconductor nanocomposites: (a) the conventional structure, (b) the core–shell structure, (c) the yolk–shell structure, (d) the Janus structure, (e) the array structure, and (f) the multijunction structure. Reprinted from [94]. CC BY 4.0.

Photocatalysts have been reported to be synthesized in nanoparticle form from materials such as carbon, metals, ceramics, semiconductors, polymers, and lipids [95]. Photocatalytic materials can be produced from a single component or they may be composite materials. The formation of composites offers many merits and is therefore the aim of many researchers [96, 97]. Nanocomposite photocatalysts such as binary [98], ternary [99] and quaternary [100] composites have been synthesized. The type of heterojunction formed in a photocatalyst is also used as basis for its categorization. Depending on the type of heterojunction, photocatalysts are classified as type I, type II, S type, or Z type [101]. Further, based on the constituent forming the major portion of the photocatalyst, they are classified as metal oxide based [102], polymer based [103] or carbon based [104]. Another category of photocatalyst is based on morphology; it includes photocatalysts in the form of sheets[105], spheres [106], rods[107], cubes [108], or flowers [109]. Other morphological forms have been observed which are considered to be important for categorizing photocatalysts. Thus, there is a wide variety of photocatalytic materials. The synthesis of novel materials is improving the performance of photocatalysts in various fields of application [110].

4.3.3 Desirable properties of photocatalysts

Photocatalysts are expected to possess certain properties related to charge transfer, activity, recyclability, and efficiency, as these properties influence the performance of photocatalysts to a great extent. Photocatalytic materials possess the ability to interact with photons, facilitating the generation of electron–hole pairs. A photocatalyst's ability to interact with photons is pivotal in determining its photocatalytic behavior. The photogenerated charges further react with pollutant species, thereby degrading them. A photocatalyst is thus required to be efficient in generating electrons and holes on exposure to photons. This interaction creates photogenerated charges and is hence given a lot of attention. Once applied for degradation purposes, photocatalysts may suffer from the problem of recombination. The goal of designing a photocatalyst is to prevent the recombination of electrons and holes. Such designs involve incorporating components that enhance the photocatalytic efficacy of the nanocomposites, which is achieved through optimization methodologies [111–114].

The improvement of photocatalytic activity is also related to structural properties. Researchers have demonstrated that structures with large surface areas [115] and small particle sizes [116] are quite useful for enhancing photocatalytic activity. Photocatalytic efficacy is observed to be influenced by morphological alterations [117]. In particular, the size of the nanoparticles and their surface area have been found to affect photocatalytic operation. In this regard, the synthesis routes adopted for making photocatalysts play a crucial role. In fact, many studies have reported variations in photocatalytic performance due to variations in process parameters [118, 119]. Photocatalytic activity is also dependent on the synthesis route. For instance, compared to conventional methods, the ultrasound-assisted method leads to a more uniform distribution of nanomaterials [120]. Photocatalysts need to be stable and strong until they reach the end of their operational life. Corrosion or wear of photocatalysts reduces their efficiency of operation. Therefore, photocatalytic operations must check the recyclability of the catalyst. It is desirable for catalysts to remain stable over the maximum number of runs [121].

Among photocatalysts, TiO_2-based catalysts are extensively used due to their efficiency in photocatalysis and their nontoxic nature. The principal limitation of TiO_2 photocatalysts is their efficacy, which is maximized under UV illumination. Efforts have been made to prepare visible-light active photocatalyst materials, as this helps in utilizing the energy of sunlight or visible light [111]. An example of this approach is to layer the photocatalyst with graphene-based material. Such nanocomposites have been found to exhibit enhanced photocatalytic properties. Graphene is known to have excellent physical, electrical, mechanical, and thermal properties, as represented in figure 4.13 [111].

In conclusion, properties such as charge carrier mobility, strength, efficiency of operation, structural properties, activity in visible light, and recyclability are the pillars of photocatalyst performance. Due to these properties, photocatalysts display

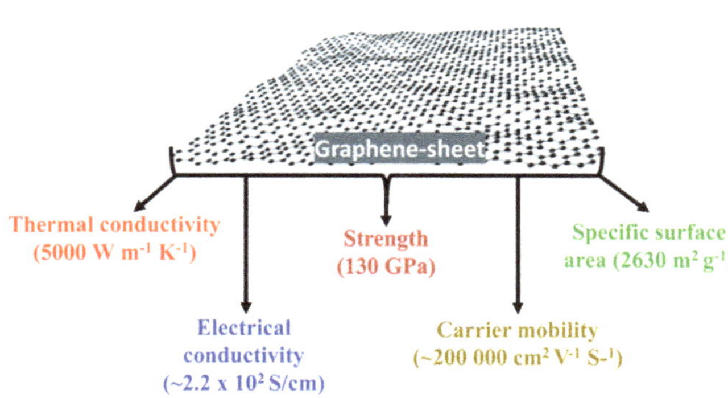

Figure 4.13. The sheet-like structure of graphene and its properties. Reproduced from [17] with permission from The Royal Society of Chemistry.

good performance in their respective applications. The benefit of photocatalytic applications is the strong degradation pathways created by room-temperature redox reactions, which are favored by photocatalysts with suitable properties.

4.3.4 How do photocatalysts work?

As mentioned in the previous section, photocatalysts generate an electron–hole pair upon interacting with a photon. The reason for the formation of this electron–hole pair is the bandgap of a material. In metals, the valence and conduction bands overlap, whereas in insulators they are separated by a large distance. Only semiconductors have a bandgap suitable for photocatalysis. In nanomaterials in which the bandgap does not fall in a range that meets the definition of a semiconductor, photocatalysis is ineffective. Therefore, the bandgap is modified by the creation of suitable composites to improve the performance of nanocomposite photocatalysts. In figure 4.14, the bandgap values of various semiconductor nanomaterials are shown. It should be noted that the performance of a photocatalyst depends on the bandgap values. Photocatalysts are active in the UV or visible range, depending on these values.

For photocatalytic degradation, we are interested in the reduction and oxidation of a pollutant. Hence, a study is necessary to perform photocatalysis via these redox reactions. Four cases are possible, based on the redox potential value: (i) at a redox potential value that is below the conduction band edge of a semiconductor, reductive reactions are possible; (ii) if the redox potential of a substrate surpasses the energy level of the valence band edge of the photocatalyst, then oxidative reactions are possible; (iii) if the redox potential of the substrate is greater than that of the conduction band or lower than that of the valence band, no reaction is expected; (iv) conversely, when the redox potential of the substrate falls below the conduction band edge yet surpasses that of the valence band, either oxidative or reductive reactions may transpire [94]. Redox reactions create various pathways for the degradation of a substrate and determine the efficiency of the photocatalytic process. To understand these paths, a photocatalytic mechanism is proposed for the degradation process.

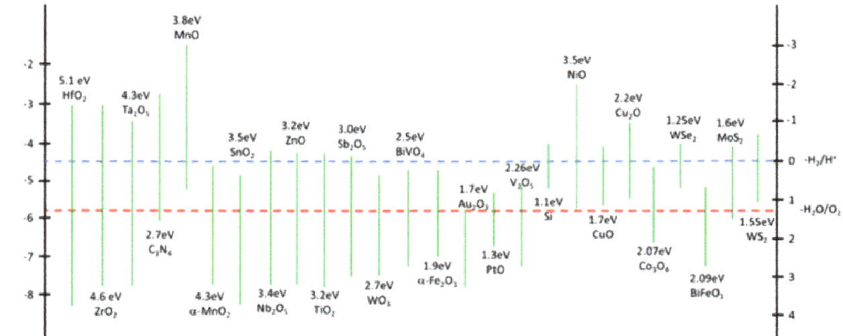

Figure 4.14. Semiconductor material bandgap values. Reprinted from [122]. CC BY 4.0.

4.3.5 Mechanisms of photocatalysis

The photocatalytic mechanism explains the stepwise pathway for the degradation of a substrate or pollutant. In general, photocatalysis involves the generation of electrons and holes which subsequently react to form other species. In particular, species with high oxidation potential values are important in photocatalysis. These powerful oxidants and their active participation are responsible for the success of photocatalysis. These oxidants mineralize pollutants into carbon dioxide and water [122]. From this study, it is clear that several species form during photocatalysis, such as electrons, holes, oxygen, oxygen anions, superoxide anions, hydroxyl radicals, and degradation products. The degradation mechanism begins with a semiconductor photocatalyst and light. It ends with degradation products. The photocatalytic mechanism thus presents the complete degradation of a pollutant in a stepwise manner, including the formation of intermediate species during the reactions.

4.3.6 Factors influencing photocatalytic activity and efficiency

In photocatalysis, photons interact with a photocatalyst, giving rise to charge carriers. The overall photocatalytic performance is therefore dependent on factors related to the source of radiation, the nature of the photocatalyst, and reactions occurring at the photocatalyst surface. As an AOP that involves light, photocatalysis is affected by many factors. The effectiveness of photocatalytic degradation for organic contaminants is contingent upon various factors, including the nature of the catalyst material, the light source, the photocatalyst's stability, the pH of the reaction, the temperature of the reaction, the surface area of the photocatalyst, and the surface morphology of the photocatalyst. These are the major factors affecting photocatalysis. As photocatalysis is a light-based operation, the light source influences photocatalysis to a great extent.

4.3.6.1 The catalyst material

Photocatalyst nanocomposites degrade pollutant species by generating electron–hole pairs. The natures and proportions of the individual component materials in a nanocomposite affect its performance. A component is added to a photocatalyst to form a nanocomposite using a doping technique. When a certain quantity of additive is included in a nanocomposite, it restricts the recombination of photo-generated charge carriers and thus improves photocatalytic operation. The selection of the dopant is a key factor in nanocomposite design. The dopant must be added in the required amount to obtain satisfactory performance from a photocatalyst. If any component material, especially the dopant, is present in excess, it does not improve photocatalytic performance [123].

Rini *et al* [109] prepared a nanocomposite photocatalyst by synthesizing zinc oxide and introducing silver doping into the zinc oxide matrix. The doping concentration of Ag was varied as follows: 0%, 2%, 4%, and 6%. An observed correlation indicated that the incremental incorporation of silver (Ag) as a dopant led to a reduction in particle size. A comparative analysis with pristine zinc oxide revealed that the Ag/ZnO nanocomposite exhibited heightened photocatalytic activity. However, at an Ag concentration of 6% in the nanocomposite, light was prevented from reaching the surface, and hence reduced photocatalytic activity was observed. This demonstrates that the optimum amount of dopant is necessary to obtain the best performance from the photocatalyst [109]. The amount of photocatalyst added to the reaction mixture also determines the photocatalytic performance. The optimum amount is decided by performing experiments with different amounts of photocatalyst. Any increase in the amount beyond the optimum value increases turbidity and reduces light penetration. Hence, it is preferable to use the optimum amount of photocatalyst [124].

4.3.6.2 The light source

The photocatalyst can be activated by a UV source, a simulated sunlight lamp, or sunlight. The photocatalytic activity is influenced by these sources. When experiments are carried out under natural light, variation due to the light is observed. Usually, the highest degradation efficiencies are observed when light intensity is at a maximum and vice versa. In order to overcome the problem of low-intensity operation, artificial light or simulated light is preferred [125]. Depending on the light intensity used, the photocatalytic degradation performance varies [126]. Lal *et al* [127] prepared Nd-doped TiO_2 and applied it to photodegrade MB using two different sources, namely UV and sunlight. The experiments carried out showed that the degradation efficiency was dependent on the source. The efficiency of Nd-doped TiO_2 was found to be 99.11% in ultraviolet light and 96.42% in solar light [127].

4.3.6.3 The photocatalyst's stability

The loss of the photocatalyst and its stability are major concerns for any photocatalytic operation. A photocatalyst must remain stable until the end of the operation and for the maximum number of runs. The wear of the photocatalysts can be avoided by designing it to remain stable for the maximum number of runs. In order to recover the photocatalyst and put it back into experimental use, recovery should be easy to perform [128]. In order to ease photocatalyst recovery, a magnetically recoverable photocatalyst has been designed [129]. This reusability test is used to determine the structural integrity of a photocatalyst [112].

4.3.6.4 The reaction conditions (pH)

Parameters related to the reaction mixture, such as temperature and pH, can change the photocatalytic operation. The pH-dependent translocation of pollutants from the reaction medium to the surface of the photocatalyst is observed in specific instances, notably with regard to dyes. In the investigation of the impact of pH on photocatalysis, it is essential to take into account the point of zero charge value [124]. Another important category is the type of reaction mixture, which is decided by the type of pollutant. A given photocatalyst may show different behavior towards different pollutants. For instance, a study designed and applied CdS–Ce/MOF nanocomposites for the photoreduction of chromium and the photocatalytic degradation of crystal violet. The nanocomposite reduced 57% of the crystal violet within a 120 min timeframe, and a 95% decrease in chromium (VI) concentration was achieved within 60 min [130]. This shows that a photocatalyst can exhibit selectivity for the degradation of a pollutant.

4.3.6.5 The surface area and morphology of the photocatalyst

The surface area is one of the characteristic properties of a nanomaterial. It is desirable for a large surface area to be available for reactions. The nanocomposite's surface area thus determines its performance. For a nanocomposite with a large surface area, there are more chances for the pollutant species to reach active sites and pollutant removal is thereby improved. For example, in research carried out by Babu *et al*, it was observed that nanomaterials that had a sheet-like morphology enhanced the photocatalytic performance. Nanosheets were prepared from BiOCl and Bi_2O_3 nanofibers and it was observed that a change in morphology resulted in a change in the photocatalytic performance [117]. In another study, a novel g-C_3N_4/$PrFe_3O_4$ nanocomposite photocatalyst was synthesized. Compared to pure $PrFe_3O_4$, the g-C_3N_4/$PrFeO_3$ nanocomposite showed improved photocatalytic activity, which was attributed to its large surface area [115]. Lellala [131] reported the creation of a mesoporous material with a high surface area composed of sulfurous carbon disc graphene oxide nanosheets. The mesoporous structure was characterized by a substantial surface area of 1050 $m^2\ g^{-1}$ and a pore size of 10 nm. The nanocomposite exhibited remarkable photocatalytic power in the degradation of brilliant yellow dye [131]. However, it is necessary to use the optimum quantities of the nanocomposite components. Only at such optimum levels is the morphology improved and the desired performance obtained [113].

4.4 Nanocomposites for photocatalytic degradation

Photocatalysis was introduced decades ago. The use of TiO_2 as a photocatalyst is widely accepted due to its properties. As a conventional photocatalyst, TiO_2 faces problems due to its inactivity in visible light. This led the scientific community to think in terms of modifying the components of photocatalysts to change their properties and performance. Designing materials while keeping in mind their properties and performance is at the core of these engineered nanocomposites. The main objective of any nanocomposite is thus to enhance performance compared to the performances of the component materials. This has paved a path for many interesting novel nanomaterials to be used for photocatalytic degradation.

Photocatalytic activity is greatly improved by creating such nanocomposites. For example, many researchers have demonstrated that introducing graphene-based material into TiO_2 results in the development of an efficient nanocomposite photocatalyst. In this case of a TiO_2 and graphene-based nanocomposite, the bandgap is changed, making the nanocomposite active in visible light [111]. It is possible to design a network of species, for example, using a metal–organic framework, and use it for the photocatalytic degradation of pollutants [132]. Morphology, as a crucial element in nanocomposite performance, is the focus of many researchers. A change in surface area is observed as a result of incorporating graphene-based components [113]. Introducing the right amount of a component into a nanocomposite affects its photocatalytic performance. This opens the door to the synthesis of nanocomposites by modifying their composition [124]. Sometimes, a layered structure is synthesized to improve its photocatalytic performance. The efficiency of a photocatalyst is impeded by the challenges related to electron–hole recombination. However, by introducing a suitable component, this recombination of electrons and holes can be suppressed. This helps to enhance the photocatalytic properties of the individual materials [112].

Photocatalysts that are applied in powdered form suffer from the problem of losses during each cycle. This problem can be overcome to a great extent by creating nanocomposite photocatalysts which are recoverable at the end of each operation. Such a nanocomposite design involves creating a stable photocatalyst that is easily recovered at the end of each run. It then becomes possible to reuse this photocatalyst in the following run [121]. Photocatalytic behavior can be improved by creating a photocatalyst that increases surface plasmons. The incorporation of metal ions, such as Ag ions, results in this type of surface plasmon effect [133].

Pollutants present in wastewater that are challenging to treat by conventional routes can be efficiently treated using nanocomposite photocatalysts. So far, nanocomposite photocatalysts have been applied for the degradation of dyes, pharmaceutical products[134], insecticides [135, 136], pesticides[137–140], and heavy metals [131, 141, 142]; they have also been used to remediate air [143] and soil [144]. Previous sections have already described that an enhancement in photocatalytic behavior is observed due to the application of nanocomposites. There are many merits of the use of nanocomposites. Nanocomposites that are designed with a tailored approach exhibit improved behavior compared to those of their individual components. The incorporation of graphene-based material into nanocomposites improves charge mobility and hence the photocatalytic performance. Sometimes, surface plasmon resonance is created by the presence of certain materials. Nanocomposites are also used in the design of engineered-bandgap materials. Many nanocomposites can make materials active under visible light. A reduction in recombination is also observed in nanocomposites. It is possible to choose component materials from a vast array of materials and design novel materials by varying the synthesis route.

4.5 Challenges and future perspectives

Conventional methods such as sedimentation, aeration, activated sludge processes, etc. are available for the treatment of effluent. Although their efficiency in the removal of pollutants is questionable, they are widely accepted due to their low cost and established status. Hence, the first challenge in the application of photocatalytic operations is scalability. The research literature describes many photocatalysts that have been synthesized and applied for the degradation of organic pollutants. Some photocatalysts have also been applied to the real industrial effluents. The application of photocatalysts to real industrial effluents is a challenge due to the presence of many contaminants in addition to the target pollutant in real industrial effluents.

For instance, textile industry effluents vary significantly in terms of their pH values, colors, metallic elements, and chemical oxygen demand. Such effluents make it difficult to streamline the treatment method [145]. Photocatalysis is dependent on light as a source of photons. For this reason, visible-light-activated photocatalysis is challenging due to variations in sunlight. The photocatalyst does not function efficiently if the sunlight is of very low intensity. On the other hand, if an artificial light is used, it overcomes the problem of the variability of light but adds to the operational costs. It is important to study the lamp properties such as diameter, length, intensity, current, voltage, and lifespan [121]. There is a need to explore new light sources such as light-emitting diodes for photocatalysis. LEDs have many merits; for example, they are free from mercury, they have longer lifespans and smaller sizes, and are physically robust. Their ability to act as a light source for photocatalysis will make it easier to carry out photocatalytic operations under reduced light and help to reduce the burden of power requirements [146].

The large-scale synthesis of nanocomposites is a major challenge which affects the use of these materials. The synthesis processes that are successful on a lab scale need to be analyzed for their use in large-scale applications [147]. The properties of nanocomposites change with changes in the method. For example, in a comparative study of synthesis methods for nanocomposite ZnO/TiO_2, two methods were used for synthesis, namely the conventional route and the ultrasound-assisted route. From the study, it was clear that the nanocomposite properties were greatly changed by the method. The nanocomposite prepared using the ultrasound method showed improved properties [148]. A method which is able to prepare the nanocomposite in an eco-friendly, cost-effective, and safe way is desirable.

A nanocomposite is a tiny entity with endless possibilities. It is crucial to evaluate the components of a nanocomposite before combining them to form a nanocomposite. The individual behaviors of the components and their behaviors in the composite are important [149]. As different tailor-made nanocomposites are synthesized, it is essential to engineer the bandgaps of these materials. Several sophisticated tools are required to synthesize and confirm the preparation of these materials [150]. A vast array of nanomaterials is available for use as photocatalysts. Most studies have presented the degradation of a pollutant by a nanomaterial. Only a few studies have reported information about the degradation products and their effect on the environment. The risks associated with intermediate products need equal attention. The toxicity profiles of nanomaterials should also be studied, so that no serious harm to the environment takes place if nanoparticles are released. Further, a detailed lifecycle assessment of nanocomposites in a particular application is also essential for their wide acceptance [151, 152].

Waste generation is a problem for the environment. Therefore, efforts are being made to use such materials as starting materials for nanocomposites. One study reported the utilization of fish waste for the preparation of a nanocomposite film. In this study, corn starch, fish scale, turmeric extract, and zinc oxide nanoparticles were used to prepare a nanocomposite film. The film displayed antimicrobial properties [153]. Sheikhy *et al* [154] reported the preparation of nanocomposites from polymer wastes. In their study, waste low-density polyethylene (LDPE), high-density polyethylene (HDPE) and nano clay were processed to produce nanocomposites [154]. Fatimah *et al* [155] prepared nanocomposites from iron rust waste using a hydrothermal method. $Fe_3O_4/NiFe_2O_4$ nanocomposites were prepared and applied for the photocatalytic degradation of methyl violet dye. The nanocomposites exhibited a degradation efficiency of more than 95% [155]. Such studies present a combined approach to the treatment of waste and the synthesis of the nanocomposite.

It is a challenging task to synthesize a nanomaterial with the desired properties using a scalable approach. In an attempt to solve this issue, Suryawanshi *et al* [156] designed a micro-reactor to produce small platinum nanoparticles capped with polyvinylpyrrolidone at a size of 5 nm. The formation of the nanoparticles was seen to be influenced by the flow of reactants [156].

Recently, Samraj *et al* [132] prepared a copper MFO–zinc tungstate heterojunction and applied it for the sonophotocatalytic degradation of tetracycline. The combined application of photocatalysis and ultrasound produced high degradation results. Such a combined or hybrid approach opens possibilities for the degradation of pollutants which are not efficiently degraded by individual methods [132].

Another interesting approach is to prepare a nanocomposite mixed matrix membrane. For example, in a study, ZIF-8 decorated cellulose acetate membranes were prepared and applied for the degradation of dyes and real textile effluent. The membrane was found to be nearly 85% efficient in removing the dyes crystal violet, acid red 13, and reactive black 5. The membrane was also found to be efficient in removing chemical oxygen demand and total organic carbon, reaching efficiencies close to 70% and 80%, respectively. Heavy metals such as lead, chromium, and cadmium were also removed by the membrane [157]. The integration of nanocomposites into membranes represents an effective strategy for the remediation of contaminated water. When a photocatalyst is prepared and applied in a powdered form, its loss in each stage is unavoidable. This problem can be solved by applying an immobilized photocatalyst, designing a photocatalytic membrane, or by synthesizing a magnetically active photocatalyst material. Effluents are rich in chemicals, and it is possible to recover this energy. A new approach is to make use of a photocatalytic fuel cell; in this way, photocatalysis and energy generation are achieved simultaneously. Such designs recover energy from wastewater while reducing its pollutant load [158].

4.6 Conclusions

The chapter provided an exhaustive examination of the evolution of nanocomposite photocatalysts in the context of degrading diverse organic pollutants. In addition, this chapter explored contemporary challenges and progresses within the domain of photocatalytic degradation. Photocatalysis is part of the family of AOPs. The successful degradation of contaminants is influenced by many factors, such as the mobility of the charge carriers, the bandgap of the photocatalyst, the process conditions, the irradiation type, and the nature of the pollutant. Therefore, the nanocomposite photocatalyst design process includes a critical study of all these factors. The effective photocatalytic decomposition of pollutants, including dyes, insecticides, pesticides, and pharmaceutical compounds has been highlighted and demonstrated by many researchers, indicating the potential of photocatalysts to degrade these substances. Some researchers have reported the degradation of real industrial effluent by combining photocatalysis with other methods. Such studies present multiple options that can be used in combination with photocatalysis for the degradation of pollutants. One such approach is to combine photocatalysis and membrane technology. Since photocatalysis is a light-dependent operation, topics such as the light source, its intensity, its penetration, and its arrangement deserve attention. A major factor in the large-scale applicability of any method is its cost. Light-efficient operation achieved using sunlight or other inexpensive light sources, such as LEDs, can reduce operational costs. While much emphasis has been given to sunlight-based operations, there is need to explore other light sources as well. In the research literature, it can be seen found that many studies have focussed on various pollutants and expanding the application of photocatalysts to a wide range of industrial contaminants. However, it is necessary to examine the details of the selective action of a photocatalyst for a particular pollutant. Such studies will open up new possibilities and degradation pathways for existing contaminants and contaminants with similar profiles. In the majority of photocatalytic processes, much emphasis has been given to degradation and this limits the application of photocatalysis to degradation alone. However, photocatalysis can degrade pollutants and simultaneously generate energy. Operations that combine photocatalysis and energy generation have many merits. Another interesting approach is to design a photocatalyst from a waste material. The waste material serves as a precursor for the synthesis. Such operations can reduce the cost of synthesis. Considering all the issues discussed in this chapter related to the synthesis of nanocomposite photocatalysts and their applications, we should anticipate the integration of cutting-edge photocatalysts with the latest methodologies in surface chemistry and materials science, leading to a new generation of catalytic entities and nano-engineering which will make photocatalysts cost-effective, scalable, and green.

References

[1] Mukherjee J, Lodh B K, Sharma R, Mahata N, Shah M P, Mandal S, Ghanta S and Bhunia B 2023 Advanced oxidation process for the treatment of industrial wastewater: a review on strategies, mechanisms, bottlenecks and prospects *Chemosphere* **345** 140473

[2] Babu Ponnusami A, Sinha S, Ashokan H, V Paul M, Hariharan S P, Arun J, Gopinath K P, Hoang Le Q and Pugazhendhi A 2023 Advanced oxidation process (AOP) combined biological process for wastewater treatment: a review on advancements, feasibility and practicability of combined techniques *Environ. Res.* **237** 116944

[3] Liu H, Li X, Zhang X, Coulon F and Wang C 2023 Harnessing the power of natural minerals: a comprehensive review of their application as heterogeneous catalysts in advanced oxidation processes for organic pollutant degradation *Chemosphere* **337** 139404
[4] Wang D, Xing Y, Li J, Dong F, Cheng H, He Z, Wang L, Giannakis S, Song S and Ma J 2023 Degradation of odor compounds in drinking water by ozone and ozone-based advanced oxidation processes: a review *ACS ES&T Water* **3** 3452–73
[5] Pattnaik A, Sahu J N, Poonia A K and Ghosh P 2023 Current perspective of nano-engineered metal oxide based photocatalysts in advanced oxidation processes for degradation of organic pollutants in wastewater *Chem. Eng. Res. Des.* **190** 667–86
[6] Navidpour A H, Abbasi S, Li D, Mojiri A and Zhou J L 2023 Investigation of advanced oxidation process in the presence of TiO_2 semiconductor as photocatalyst: property, principle, kinetic analysis, and photocatalytic activity *Catalysts* **13** 232
[7] Amulya M A S, Nagaswarupa H P, Kumar M R A, Ravikumar C R, Prashantha S C and Kusuma K B 2020 Sonochemical synthesis of $NiFe_2O_4$ nanoparticles: characterization and their photocatalytic and electrochemical applications *Appl. Surf. Sci. Adv.* **1** 100023
[8] Zhang Q, Uchaker E, Candelaria S L and Cao G 2013 Nanomaterials for energy conversion and storage *Chem. Soc. Rev.* **42** 3127
[9] Chauhan H A, Rafatullah M, Ahmed Ali K, Siddiqui M R, Khan M A and Alshareef S A 2021 Metal-based nanocomposite materials for efficient photocatalytic degradation of phenanthrene from aqueous solutions *Polymers (Basel)* **13** 2374
[10] Boruah P K, Borthakur P and Das M R 2019 Magnetic metal/metal oxide nanoparticles and nanocomposite materials for water purification *Nanoscale Materials in Water Purification* (Amsterdam: Elsevier) 473–503
[11] Mavrikos A *et al* 2022 Synthesis of Zn/Cu metal ion modified natural palygorskite clay–TiO_2 nanocomposites for the photocatalytic outdoor and indoor air purification *J. Photochem. Photobiol. A Chem.* **423** 113568
[12] Pandey N, Shukla S K and Singh N B 2017 Water purification by polymer nanocomposites: an overview *Nanocomposites* **3** 47–66
[13] Xia C, Li X, Wu Y, Suharti S, Unpaprom Y and Pugazhendhi A 2023 A review on pollutants remediation competence of nanocomposites on contaminated water *Environ. Res.* **222** 115318
[14] Kuvarega A T and Mamba B B 2017 TiO_2-based photocatalysis: toward visible light-responsive photocatalysts through doping and fabrication of carbon-based nanocomposites *Crit. Rev. Solid State Mater. Sci.* **42** 295–346
[15] Gopannagari M, Kumar D P, Park H, Kim E H, Bhavani P, Reddy D A and Kim T K 2018 Influence of surface-functionalized multi-walled carbon nanotubes on CdS nano-hybrids for effective photocatalytic hydrogen production *Appl. Catal.* B **236** 294–303
[16] Bathla A, Vikrant K, Kukkar D and Kim K-H 2022 Photocatalytic degradation of gaseous benzene using metal oxide nanocomposites *Adv. Colloid Interface Sci.* **305** 102696
[17] Khan M E 2021 State-of-the-art developments in carbon-based metal nanocomposites as a catalyst: photocatalysis *Nanoscale Adv.* **3** 1887–900
[18] Akakuru O U, Iqbal Z M and Wu A 2020 TiO2 Nanoparticles: Properties and Applications *TiO2 Nanoparticles: Applications in Nanobiotechnology and Nanomedicine* (New York: Wiley) 1 1–66
[19] Xu F 2018 Review of analytical studies on TiO_2 nanoparticles and particle aggregation, coagulation, flocculation, sedimentation, stabilization *Chemosphere* **212** 662–77

[20] Upadhyay G K, Rajput J K, Pathak T K, Kumar V and Purohit L P 2019 Synthesis of ZnO:TiO_2 nanocomposites for photocatalyst application in visible light *Vacuum* **160** 154–63

[21] Bethi B, Sonawane S H, Rohit G S, Holkar C R, Pinjari D V, Bhanvase B A and Pandit A B 2016 Investigation of TiO_2 photocatalyst performance for decolorization in the presence of hydrodynamic cavitation as hybrid AOP *Ultrason. Sonochem.* **28** 150–60

[22] Chand P and Singh V 2020 Enhanced visible-light photocatalytic activity of samarium-doped zinc oxide nanostructures *J. Rare Earths* **38** 29–38

[23] Qutub N, Singh P, Sabir S, Sagadevan S and Oh W-C 2022 Enhanced photocatalytic degradation of acid blue dye using CdS/TiO_2 nanocomposite *Sci. Rep.* **12** 5759

[24] Karimi-Maleh H, Kumar B G, Rajendran S, Qin J, Vadivel S, Durgalakshmi D, Gracia F, Soto-Moscoso M, Orooji Y and Karimi F 2020 Tuning of metal oxides photocatalytic performance using Ag nanoparticles integration *J. Mol. Liq.* **314** 113588

[25] Azizi-Lalabadi M, Alizadeh-Sani M, Divband B, Ehsani A and McClements D J 2020 Nanocomposite films consisting of functional nanoparticles (TiO_2 and ZnO) embedded in 4A-Zeolite and mixed polymer matrices (gelatin and polyvinyl alcohol) *Food Res. Int.* **137** 109716

[26] Malesic-Eleftheriadou N, Evgenidou E, Kyzas G Z, Bikiaris D N and Lambropoulou D A 2019 Removal of antibiotics in aqueous media by using new synthesized bio-based poly (ethylene terephthalate)–TiO_2 photocatalysts *Chemosphere* **234** 746–55

[27] Feizpoor S, Habibi-Yangjeh A, Yubuta K and Vadivel S 2019 Fabrication of TiO_2/$CoMoO_4$/PANI nanocomposites with enhanced photocatalytic performances for removal of organic and inorganic pollutants under visible light *Mater. Chem. Phys.* **224** 10–21

[28] Faisal M, Jalalah M, Harraz F A, El-Toni A M, Labis J P and Al-Assiri M S 2021 A novel Ag/PANI/$ZnTiO_3$ ternary nanocomposite as a highly efficient visible-light-driven photocatalyst *Sep. Purif. Technol.* **256** 117847

[29] Kumar A, Raorane C J, Syed A, Bahkali A H, Elgorban A M, Raj V and Kim S C 2023 Synthesis of TiO_2, TiO_2/PAni, TiO_2/PAni/GO nanocomposites and photodegradation of anionic dyes Rose Bengal and thymol blue in visible light *Environ. Res.* **216** 114741

[30] Munyengabe A, Ndibewu P P, Sibali L L and Ngobeni P 2022 Polymeric nanocomposite materials for photocatalytic detoxification of polycyclic aromatic hydrocarbons in aquatic environments-a review *Res. Eng.* **15** 100530

[31] Ningthoujam R, Singh Y D, Babu P J, Tirkey A, Pradhan S and Sarma M 2022 Nanocatalyst in remediating environmental pollutants *Chem. Phys. Impact* **4** 100064

[32] Khakhal H R, Kumar S, Patidar D, Kumar S, Vats V S, Dalela B, Alvi P A, Leel N S and Dalela S 2023 Correlation of oxygen defects, oxide-ion conductivity and dielectric relaxation to electronic structure and room temperature ferromagnetic properties of Yb^{3+} doped CeO_2 nanoparticles *Mater. Sci. Eng.:* B **297** 116675

[33] Bakkiyaraj R, Bharath G, Hasini Ramsait K, Abdel-Wahab A, Alsharaeh E H, Chen S-M and Balakrishnan M 2016 Solution combustion synthesis and physico-chemical properties of ultrafine CeO_2 nanoparticles and their photocatalytic activity *RSC Adv.* **6** 51238–45

[34] Zinatloo-Ajabshir S, Heidari-Asil S A and Salavati-Niasari M 2021 Simple and eco-friendly synthesis of recoverable zinc cobalt oxide-based ceramic nanostructure as high-performance photocatalyst for enhanced photocatalytic removal of organic contamination under solar light *Sep. Purif. Technol.* **267** 118667

[35] Mahdavi K, Zinatloo-Ajabshir S, Yousif Q A and Salavati-Niasari M 2022 Enhanced photocatalytic degradation of toxic contaminants using Dy_2O_3–SiO_2 ceramic nanostructured materials fabricated by a new, simple and rapid sonochemical approach *Ultrason. Sonochem.* **82** 105892

[36] Kumar K, Kumar R, Kaushal S, Thakur N, Umar A, Akbar S, Ibrahim A A and Baskoutas S 2023 Biomass waste-derived carbon materials for sustainable remediation of polluted environment: a comprehensive review *Chemosphere* **345** 140419

[37] Ahmed M K, Shalan A E, Afifi M, El-Desoky M M and Lanceros-Méndez S 2021 Silver-doped cadmium selenide/graphene oxide-filled cellulose acetate nanocomposites for photocatalytic degradation of malachite green toward wastewater treatment *ACS Omega* **6** 23129–38

[38] Aragaw B A and Dagnaw A 1970 Copper/reduced graphene oxide nanocomposite for high performance photocatalytic methylene blue dye degradation *Ethiop. J. Sci. Technol.* **12** 125–37

[39] Beura R, Pachaiappan R and Paramasivam T 2021 Photocatalytic degradation studies of organic dyes over novel Ag-loaded ZnO-graphene hybrid nanocomposites *J. Phys. Chem. Solids* **148** 109689

[40] Gu Y, Xing M and Zhang J 2014 Synthesis and photocatalytic activity of graphene based doped TiO_2 nanocomposites *Appl. Surf. Sci.* **319** 8–15

[41] Li C, Wang B, Zhang F, Song N, Liu G, Wang C and Zhong S 2020 Performance of Ag/BiOBr/GO composite photocatalyst for visible-light-driven dye pollutants degradation *J. Mater. Res. Technol.* **9** 610–21

[42] Fulzele N N, Bhanvase B A and Pandharipande S L 2022 Sonochemically prepared rGO/Ag_3PO_4/CeO_2 nanocomposite photocatalyst for effective visible light photocatalytic degradation of methylene dye and its prediction with ANN modeling *Mater. Chem. Phys.* **292** 126809

[43] Ardani M R, Pang A L, Pal U, Zheng R, Arsad A, Hamzah A A and Ahmadipour M 2022 Ultrasonic-assisted polyaniline-multiwall carbon nanotube photocatalyst for efficient photodegradation of organic pollutants *J. Water Process. Eng.* **46** 102557

[44] Sadeghi Rad T, Khataee A, Sadeghi Rad S, Arefi-Oskoui S, Gengec E, Kobya M and Yoon Y 2022 Zinc-chromium layered double hydroxides anchored on carbon nanotube and biochar for ultrasound-assisted photocatalysis of rifampicin *Ultrason. Sonochem.* **82** 105875

[45] Zhang X, Wang Q, Zou L-H and You J-W 2016 Facile fabrication of titanium dioxide/fullerene nanocomposite and its enhanced visible photocatalytic activity *J. Colloid Interface Sci.* **466** 56–61

[46] Abid N, Khan A M, Shujait S, Chaudhary K, Ikram M, Imran M, Haider J, Khan M, Khan Q and Maqbool M 2022 Synthesis of nanomaterials using various top-down and bottom-up approaches, influencing factors, advantages, and disadvantages: a review *Adv. Colloid Interface Sci.* **300** 102597

[47] Thambiliyagodage C, Mirihana S, Wijesekera R, Madusanka D S, Kandanapitiye M and Bakker M 2021 Fabrication of Fe_2TiO_5/TiO_2 binary nanocomposite from natural ilmenite and their photocatalytic activity under solar energy *Curr. Res. Green Sustain. Chem.* **4** 100156

[48] Pathak D, Sharma A, Sharma D P and Kumar V 2023 A review on electrospun nanofibers for photocatalysis: upcoming technology for energy and environmental remediation applications *Appl. Surf. Sci. Adv.* **18** 100471

[49] Wu C 2014 Synthesis of Ag 2 CO 3 /ZnO nanocomposite with visible light-driven photocatalytic activity *Mater. Lett.* **136** 262–4

[50] Haghighatzadeh A, Hosseini M, Mazinani B and Shokouhimehr M 2019 Improved photocatalytic activity of ZnO-TiO 2 nanocomposite catalysts by modulating TiO 2 thickness *Mater. Res. Express* **6** 115060

[51] Haounati R *et al* 2021 Design of direct Z-scheme superb magnetic nanocomposite photocatalyst Fe_3O_4/Ag_3PO_4@Sep for hazardous dye degradation *Sep. Purif. Technol.* **277** 119399

[52] Kumar M, Singh R, Khajuria H and Sheikh H N 2017 Facile hydrothermal synthesis of nanocomposites of nitrogen doped graphene with metal molybdates (NG-$MMoO_4$) (M = Mn, Co, and Ni) for enhanced photodegradation of methylene blue *J. Mater. Sci., Mater. Electron.* **28** 9423–34

[53] Younas U, Ahmad A, Islam A, Ali F, Pervaiz M, Saleem A, Waseem M, Muteb Aljuwayid A, Habila M A and Raza Naqvi S 2023 Fabrication of a novel nanocomposite ($TiO_2/WO_3/V_2O_5$) by hydrothermal method as catalyst for hazardous waste treatment *Fuel* **349** 128668

[54] Parasuraman B, Kandasamy B, Murugan I, Alsalhi M S, Asemi N, Thangavelu P and Perumal S 2023 Designing the heterostructured $FeWO_4/FeS_2$ nanocomposites for an enhanced photocatalytic organic dye degradation *Chemosphere* **334** 138979

[55] Shibu M C, Benoy M D, Kumar G S, Duraimurugan J, Vasudevan V, Shkir M and AL-Otaibi O 2023 Hydrothermal-assisted synthesis and characterization of MWCNT/copper oxide nanocomposite for the photodegradation of methyl orange under direct sunlight *Diam. Relat. Mater.* **134** 109778

[56] Zinatloo-Ajabshir S, Heidari-Asil S A and Salavati-Niasari M 2021 Recyclable magnetic $ZnCo_2O_4$-based ceramic nanostructure materials fabricated by simple sonochemical route for effective sunlight-driven photocatalytic degradation of organic pollution *Ceram. Int.* **47** 8959–72

[57] Warsi A-Z, Hussien O K, Iftikhar A, Aziz F, Alhashmialameer D, Mahmoud S F, Warsi M F and Saleh D I 2022 Co-precipitation assisted preparation of Ag_2O, CuO and Ag_2O/CuO nanocomposite: characterization and improved solar irradiated degradation of colored and colourless organic effluents *Ceram. Int.* **48** 19056–67

[58] Pudukudy M, Jia Q, Yuan J, Megala S, Rajendran R and Shan S 2020 Influence of CeO_2 loading on the structural, textural, optical and photocatalytic properties of single-pot sol–gel derived ultrafine CeO_2/TiO_2 nanocomposites for the efficient degradation of tetracycline under visible light irradiation *Mater. Sci. Semicond. Process.* **108** 104891

[59] Bai N, Liu X, Li Z, Ke X, Zhang K and Wu Q 2021 High-efficiency TiO_2/ZnO nanocomposites photocatalysts by sol–gel and hydrothermal methods *J. Solgel Sci. Technol.* **99** 92–100

[60] Akhter P, Nawaz S, Shafiq I, Nazir A, Shafique S, Jamil F, Park Y-K and Hussain M 2023 Efficient visible light assisted photocatalysis using ZnO/TiO_2 nanocomposites *Mol. Catal.* **535** 112896

[61] Khilji M-U-N, Nahyoon N A, Mehdi M, Thebo K H, Mahar N, Memon A A, Memon N and Hussain N 2023 Synthesis of novel visible light driven MgO@GO nanocomposite photocatalyst for degradation of Rhodamine 6G *Opt. Mater.* **135** 113260

[62] Alsalme A, AlFawaz A, Glal A H, Abdel Messih M F, Soltan A and Ahmed M A 2023 S-scheme $AgIO_4/CeO_2$ heterojunction nanocomposite photocatalyst for degradation of rhodamine B dye *J. Photochem. Photobiol. A Chem.* **439** 114596

[63] Ur Rahman Z, Shah U, Alam A, Shah Z, Shaheen K, Bahadar Khan S and Ali Khan S 2023 Photocatalytic degradation of cefixime using CuO-NiO nanocomposite photocatalyst *Inorg. Chem. Commun.* **148** 110312
[64] Alajmi B M, Basaleh A S, Ismail A A and Mohamed R M 2023 Hierarchical mesoporous CuO/ZrO_2 nanocomposite photocatalyst for highly stable photoinduced desulfurization of thiophene *Surf. Interfaces* **39** 102899
[65] Xu J, Su S, Song X, Luo S, Ye S and Situ W 2023 A simple nanocomposite photocatalyst HT-rGO/TiO_2 for deoxynivalenol degradation in liquid food *Food Chem.* **408** 135228
[66] Baig U, Dastageer M A, Gondal M A and Khalil A B 2023 Photocatalytic deactivation of sulphate reducing bacteria using visible light active CuO/TiO_2 nanocomposite photocatalysts synthesized by ultrasonic processing *J. Photochem. Photobiol.*, B **242** 112698
[67] Zare A, Saadati A and Sheibani S 2023 Modification of a Z-scheme ZnO-CuO nanocomposite by Ag loading as a highly efficient visible light photocatalyst *Mater. Res. Bull.* **158** 112048
[68] Verma N, Chundawat T S, Chandra H and Vaya D 2023 An efficient time reductive photocatalytic degradation of carcinogenic dyes by TiO_2–GO nanocomposite *Mater. Res. Bull.* **158** 112043
[69] Packialakshmi J S, Albeshr M F, Alrefaei A F, Zhang F, Liu X, Selvankumar T and Mythili R 2023 Development of ZnO/SnO_2/rGO hybrid nanocomposites for effective photocatalytic degradation of toxic dye pollutants from aquatic ecosystems *Environ. Res.* **225** 115602
[70] Vivek P, Sivakumar R, Selva Esakki E and Deivanayaki S 2023 Fabrication of NiO/RGO nanocomposite for enhancing photocatalytic performance through degradation of RhB *J. Phys. Chem. Solids* **176** 111255
[71] Sabouri Z, Kazemi Oskuee R, Sabouri S, Tabrizi Hafez Moghaddas S S, Samarghandian S, Sajid Abdulabbas H and Darroudi M 2023 Phytoextract-mediated synthesis of Ag-doped ZnO–MgO–CaO nanocomposite using *Ocimum basilicum* L seeds extract as a highly efficient photocatalyst and evaluation of their biological effects *Ceram. Int.* **49** 20989–97
[72] Manda A A *et al* 2023 Fast one-pot laser-based fabrication of ZnO/TiO_2-reduced graphene oxide nanocomposite for photocatalytic applications *Opt. Laser Technol.* **160** 109105
[73] Chinnasamy C, Perumal N, Choubey A and Rajendran S 2023 Recent advancements in MXene-based nanocomposites as photocatalysts for hazardous pollutant degradation—a review *Environ. Res.* **233** 116459
[74] Purkayastha M D, Pal Majumder T, Sarkar M and Ghosh S 2022 Carrier transport and shielding properties of rod-like mesoporous TiO_2–SiO_2 nanocomposite *Radiat. Phys. Chem.* **192** 109898
[75] Tamayo-Vegas S and Lafdi K 2022 Experimental and modelling of temperature-dependent mechanical properties of CNT/polymer nanocomposites *Mater. Today Proc.* **57** 607–14
[76] Osman A, Elhakeem A, Kaytbay S and Ahmed A 2022 A comprehensive review on the thermal, electrical, and mechanical properties of graphene-based multi-functional epoxy composites *Adv. Compos. Hybrid Mater.* **5** 547–605
[77] Norizan M N, Abdullah N, Halim N A, Demon S Z N and Mohamad I S 2022 Heterojunctions of rGO/metal oxide nanocomposites as promising gas-sensing materials —a review *Nanomaterials* **12** 2278
[78] Mehta M, Chandrabose G, Krishnamurthy S, Avasthi D K and Chowdhury S 2022 Improved photoelectrochemical properties of TiO_2–graphene nanocomposites: effect of

defect induced visible light absorption and graphene conducting channel for carrier transport *Appl. Surf. Sci. Adv.* **11** 100274
[79] Williams J D and Peterson G P 2021 A review of thermal property enhancements of low-temperature nano-enhanced phase change materials *Nanomaterials* **11** 2578
[80] Alzahrani H S, Al-Sulami A I, Alsulami Q A and Rajeh A 2022 A systematic study of structural, conductivity, linear, and nonlinear optical properties of PEO/PVA-MWCNTs/ZnO nanocomposites films for optoelectronic applications *Opt. Mater.* **133** 112900
[81] Chang C-Y, Yamakata A and Tseng W J 2022 Effect of surface plasmon resonance and the heterojunction on photoelectrochemical activity of metal-loaded TiO_2 electrodes under visible light irradiation *J. Phys. Chem.* C **126** 12450–9
[82] Bethi B, Sonawane S H, Bhanvase B A and Gumfekar S P 2016 Nanomaterials-based advanced oxidation processes for wastewater treatment: a review *Chem. Engi. Process.* **109** 178–89
[83] Zhang Z, Zhang X, Porcar-Castell A, Chen J M, Ju W, Wu L, Wu Y and Zhang Y 2022 Sun-induced chlorophyll fluorescence is more strongly related to photosynthesis with hemispherical than nadir measurements: evidence from field observations and model simulations *Remote Sens. Environ.* **279** 113118
[84] Kubacka A, Caudillo-Flores U, Barba-Nieto I and Fernández-García M 2021 Towards full-spectrum photocatalysis: successful approaches and materials *Appl. Catal. A Gen.* **610** 117966
[85] Gallagher J M, Roberts B M W, Borsley S and Leigh D A 2023 Conformational selection accelerates catalysis by an organocatalytic molecular motor *Chem.*
[86] Ohtani B 2010 Photocatalysis A to Z—what we know and what we do not know in a scientific sense *J. Photochem. Photobiol.,* C **11** 157–78
[87] Petronella F, Truppi A, Ingrosso C, Placido T, Striccoli M, Curri M L, Agostiano A and Comparelli R 2017 Nanocomposite materials for photocatalytic degradation of pollutants *Catal. Today* **281** 85–100
[88] Wu H, Yang K, Wang X, Fang N, Weng P, Duan L, Zhang C, Wang X and Liu L 2023 Xenon-lamp simulated sunlight-induced photolysis of pyriclobenzuron in water: kinetics, degradation pathways, and identification of photolysis products *Ecotoxicol. Environ. Saf.* **263** 115272
[89] Łoński S, Łukowiec D, Barbusiński K, Babilas R, Szeląg B and Radoń A 2023 Flower-like magnetite nanoparticles with unfunctionalized surface as an efficient catalyst in photo-Fenton degradation of chemical dyes *Appl. Surf. Sci.* **638** 158127
[90] Shankaraiah G, Saritha P, Bhagawan D, Himabindu V and Vidyavathi S 2017 Photochemical oxidation of antibiotic gemifloxacin in aqueous solutions—a comparative study *S. Afr. J. Chem. Eng.* **24** 8–16
[91] Torres-Pinto A, Díez A M, Silva C G, Faria J L, Sanromán M Á, Silva A M T and Pazos M 2023 Photoelectrocatalytic degradation of pharmaceuticals promoted by a metal-free g-C_3N_4 catalyst *Chem. Eng. J.* **476** 146761
[92] Wu S, Jia Q and Dai W 2017 Synthesis of RGO/TiO 2 hybrid as a high performance photocatalyst *Ceram. Int.* **43** 1530–5
[93] Sang Y, Liu H and Umar A 2015 Photocatalysis from UV/Vis to near-infrared light: towards full solar-light spectrum activity *ChemCatChem.* **7** 559–73
[94] Fu Y, Li J and Li J 2019 Metal/semiconductor nanocomposites for photocatalysis: fundamentals, structures, applications and properties *Nanomaterials* **9** 359

[95] Khan I, Saeed K and Khan I 2019 Nanoparticles: properties, applications and toxicities *Arab. J. Chem.* **12** 908–31

[96] Sharma G, Kumar A, Sharma S, Naushad M, Prakash Dwivedi R, ALOthman Z A and Mola G T 2019 Novel development of nanoparticles to bimetallic nanoparticles and their composites: a review *J King Saud Univ. Sci.* **31** 257–69

[97] Raizada P, Sudhaik A and Singh P 2019 Photocatalytic water decontamination using graphene and ZnO coupled photocatalysts: a review *Mater. Sci. Energy Technol.* **2** 509–25

[98] Bresolin B-M, Sgarbossa P, Bahnemann D W and Sillanpää M 2020 $Cs_3Bi_2I_9$/g-C_3N_4 as a new binary photocatalyst for efficient visible-light photocatalytic processes *Sep. Purif. Technol.* **251** 117320

[99] Gu J, Jia H, Ma S, Ye Z, Pan J, Dong R, Zong Y and Xue J 2020 Fe_3O_4-loaded g-C_3N_4/C-layered composite as a ternary photocatalyst for tetracycline degradation *ACS Omega* **5** 30980–8

[100] Lin Y-Y, Hung J-T, Chou Y-C, Shen S-J, Wu W-T, Liu F-Y, Lin J-H and Chen C-C 2022 Synthesis of bismuth oxybromochloroiodide/graphitic carbon nitride quaternary composites (BiOxCly/BiOmBrn/BiOpIq/g-C_3N_4) enhances visible-light-driven photocatalytic activity *Catal. Commun.* **163** 106418

[101] Wang G, Lv S, Shen Y, Li W, Lin L and Li Z 2023 Advancements in heterojunction, cocatalyst, defect and morphology engineering of semiconductor oxide photocatalysts *J. Materiomics*

[102] Russo S, Muscetta M, Amato P, Venezia V, Verrillo M, Rega R, Lettieri S, Cocca M, Marotta R and Vitiello G 2024 Humic substance/metal-oxide multifunctional nanoparticles as advanced antibacterial-antimycotic agents and photocatalysts for the degradation of PLA microplastics under UVA/solar radiation *Chemosphere* **346** 140605

[103] Liu Y, Xu L, Zhang N, Wang J, Mu X and Wang Y 2022 A promoted charge separation/transfer and surface plasmon resonance effect synergistically enhanced photocatalytic performance in Cu nanoparticles and single-atom Cu supported attapulgite/polymer carbon nitride photocatalyst *Mater. Today Chem.* **26** 101250

[104] Sarwar A, Razzaq A, Zafar M, Idrees I, Rehman F and Kim W Y 2023 Copper tungstate ($CuWO_4$)/graphene quantum dots (GQDs) composite photocatalyst for enhanced degradation of phenol under visible light irradiation *Results Phys.* **45** 106253

[105] Hafeez H Y, Mohammed J, Suleiman A B, Ndikilar C E, Sa'id R S and Muhammad I 2023 Insights into hybrid TiO_2–g-C_3N_4 heterostructure composite decorated with rGO sheet: a highly efficient photocatalyst for boosted solar fuel (hydrogen) generation *Chem. Phys. Impact* **6** 100157

[106] Zhao Q *et al* 2020 Nonhydrolytic sol–gel in-situ synthesis of novel recoverable amorphous Fe_2TiO_5/C hollow spheres as visible-light driven photocatalysts *Mater. Des.* **194** 108928

[107] Wu P, Xu W, Gu X, Wang M, Liu B and Khan S 2020 Preparation of rod-shaped Bi_5O_7I as Bifunctional Material for Supercapacitors and Photcatalysts *Int. J. Electrochem. Sci.* **15** 11294–305

[108] Adekoya J A, Chibuokem M O, Masikane S and Revaprasadu N 2023 Heterostructures of Ag_2FeSnS_4 chalcogenide nanoparticles as potential photocatalysts *Sci. Afr.* **19** e01509

[109] Rini A S, Defti A P, Dewi R, Jasril and Rati Y 2023 Biosynthesis of nanoflower Ag-doped ZnO and its application as photocatalyst for Methylene blue degradation *Mater. Today Proc.* **87** 234–9

[110] Di Paola A, García-López E, Marcì G and Palmisano L 2012 A survey of photocatalytic materials for environmental remediation *J. Hazard. Mater.* **211–212** 3–29
[111] Tayebi M, Kolaei M, Tayyebi A, Masoumi Z, Belbasi Z and Lee B-K 2019 Reduced graphene oxide (RGO) on TiO_2 for an improved photoelectrochemical (PEC) and photocatalytic activity *Sol. Energy* **190** 185–94
[112] Gupta S and Subramanian V R 2014 Encapsulating $Bi_2Ti_2O_7$ (BTO) with reduced graphene oxide (RGO): an effective strategy to enhance photocatalytic and photoelectrocatalytic activity of BTO *ACS Appl. Mater. Interfaces* **6** 18597–608
[113] Moztahida M, Jang J, Nawaz M, Lim S-R and Lee D S 2019 Effect of rGO loading on Fe_3O_4: a visible light assisted catalyst material for carbamazepine degradation *Sci. Total Environ.* **667** 741–50
[114] Ben Saber N, Mezni A, Alrooqi A and Altalhi T 2021 Fabrication of efficient Au@TiO_2/rGO heterojunction nanocomposite: boosted photocatalytic activity under ultraviolet and visible light irradiation *J. Mater. Res. Technol.* **12** 2238–46
[115] Chebanenko M I, Lebedev L A, Seroglazova A S, Lobinsky A A, Gerasimov E Y, Stovpiaga E Y and Popkov V I 2023 Novel g-C_3N_4/$PrFeO_3$ nanocomposites with Z-scheme structure and superior photocatalytic activity toward visible-light-driven removal of tetracycline antibiotics *Heliyon* **9** e22038
[116] Quan Y, YiO M H N, Li Y, Myers R J and Kafizas A 2023 Influence of Bi co-catalyst particle size on the photocatalytic activity of BiOI microflowers in Bi/BiOI junctions—a mechanistic study of charge carrier behaviour *J. Photochem. Photobiol. A Chem.* **443** 114889
[117] Babu V J, Bhavatharini R S R and Ramakrishna S 2014 Bi_2O_3 and BiOCl electrospun nanosheets and morphology-dependent photocatalytic properties *RSC Adv.* **4** 29957
[118] Pirgholi-Givi G, Farjami-Shayesteh S and Azizian-Kalandaragh Y 2019 The influence of preparation parameters on the photocatalytic performance of mixed bismuth titanate-based nanostructures *Physica* B **575** 311572
[119] Zou H, Song M, Yi F, Wang X, Bian L, Li W, Zhang J, Pan N and Zeng P 2018 Effect of sintering temperature on the photocatalytic activity of carbon–Bi_2O_3–TiO_2 composite *J. Mater. Sci., Mater. Electron.* **29** 2201–8
[120] Kallawar G A, Bhanvase B A and Sathe B R 2023 Sonochemically prepared bismuth doped titanium oxide-reduced graphene oxide (Bi@TiO_2–rGO) nanocomposites for effective visible light photocatalytic degradation of malachite green *Diam. Relat. Mater.* **139** 110423
[121] Grzegórska A, Ofoegbu J C, Cervera-Gabalda L, Gómez-Polo C, Sannino D and Zielińska-Jurek A 2023 Magnetically recyclable TiO_2/MXene/$MnFe_2O_4$ photocatalyst for enhanced peroxymonosulphate-assisted photocatalytic degradation of carbamazepine and ibuprofen under simulated solar light *J. Environ. Chem. Eng.* **11** 110660
[122] Porcu S, Secci F and Ricci P C 2022 Advances in hybrid composites for photocatalytic applications: a review *Molecules* **27** 6828
[123] Irfan S, Zhuanghao Z, Li F, Chen Y-X, Liang G-X, Luo J-T and Ping F 2019 Critical review: Bismuth ferrite as an emerging visible light active nanostructured photocatalyst *J. Mater. Res. Technol.* **8** 6375–89
[124] Tegenaw A B, Yimer A A and Beyene T T 2023 Boosting the photocatalytic activity of ZnO-NPs through the incorporation of C-dot and preparation of nanocomposite materials *Heliyon* **9** e20717

[125] Yang L, Guo J, Yang T, Guo C, Zhang S, Luo S, Dai W, Li B, Luo X and Li Y 2021 Self-assembly Cu_2O nanowire arrays on Cu mesh: a solid-state, highly-efficient, and stable photocatalyst for toluene degradation under sunlight *J. Hazard. Mater.* **402** 123741

[126] Meng Y, Huang X, Wu Y, Wang X and Qian Y 2002 Kinetic study and modeling on photocatalytic degradation of para-chlorobenzoate at different light intensities *Environ. Pollut.* **117** 307–13

[127] Lal M, Sharma P and Ram C 2022 Synthesis and photocatalytic potential of Nd-doped TiO_2 under UV and solar light irradiation using a sol–gel ultrasonication method *Res. Mater.* **15** 100308

[128] Yin Z-C, Yang M, Gosavi S W, Kumar Singh A, Chauhan R and Jin J-C 2021 A 3D supramolecular Ag(I)-based coordination polymer as stable photocatalyst for dye degradation *Inorg. Chem. Commun.* **131** 108805

[129] Zhang Q, Yu L, Xu C, Zhao J, Pan H, Chen M, Xu Q and Diao G 2019 Preparation of highly efficient and magnetically recyclable Fe_3O_4@C@Ru nanocomposite for the photocatalytic degradation of methylene blue in visible light *Appl. Surf. Sci.* **483** 241–51

[130] Meng J, He M, Li F, Li T, Huang Z and Cao W 2023 Combining Ce–metal–organic framework with CdS for efficient photocatalytic removals of heavy metal ion and organic pollutant under visible and solar lights *Inorganica Chim. Acta* **557** 121701

[131] Lellala K 2021 Sulfur embedded on in-situ carbon nanodisc decorated on graphene sheets for efficient photocatalytic activity and capacitive deionization method for heavy metal removal *J. Mater. Res. Technol.* **13** 1555–66

[132] Jeyaprakash J S, Rajamani M, Bianchi C L, Ashokkumar M and Neppolian B 2023 Highly efficient ultrasound-driven Cu–MOF/$ZnWO_4$ heterostructure: an efficient visible-light photocatalyst with robust stability for complete degradation of tetracycline *Ultrason. Sonochem.* **100** 106624

[133] Leong K H, Gan B L, Ibrahim S and Saravanan P 2014 Synthesis of surface plasmon resonance (SPR) triggered Ag/TiO_2 photocatalyst for degradation of endocrine disturbing compounds *Appl. Surf. Sci.* **319** 128–35

[134] Antonopoulou M, Kosma C, Albanis T and Konstantinou I 2021 An overview of homogeneous and heterogeneous photocatalysis applications for the removal of pharmaceutical compounds from real or synthetic hospital wastewaters under lab or pilot scale *Sci. Total Environ.* **765** 144163

[135] McCormick W J, McCrudden D, Skillen N and Robertson P K J 2023 Electrochemical monitoring of the photocatalytic degradation of the insecticide emamectin benzoate using TiO_2 and ZnO materials *Appl. Catal. A Gen.* **660** 119201

[136] Zelić I E, Povijač K, Gilja V, Tomašić V and Gomzi Z 2022 Photocatalytic degradation of acetamiprid in a rotating photoreactor—determination of reactive species *Catal. Commun.* **169** 106474

[137] Li S-S, Wen L, He S-W, Xu Z, Ding L, Cheng Y-H and Chen M-L 2023 Enhanced photocatalytic degradation of imidacloprid by a simple Z-type binary heterojunction composite of long afterglow with metal-organic framework *Catal. Commun.* **183** 106775

[138] Zhang Y, Cao X, Yang Y, Guan S, Wang X, Li H, Zheng X, Zhou L, Jiang Y and Gao J 2023 Visible light assisted enzyme-photocatalytic cascade degradation of organophosphorus pesticides *Green Chem. Eng.* **4** 30–8

[139] Ayodhya D and Veerabhadram G 2019 Fabrication of Schiff base coordinated ZnS nanoparticles for enhanced photocatalytic degradation of chlorpyrifos pesticide and detection of heavy metal ions *J. Materiomics* **5** 446–54
[140] Yogesh Kumar K, Prashanth M K, Shanavaz H, Parashuram L, Alharethy F, Jeon B-H and Raghu M S 2023 Ultrasound assisted fabrication of $InVO_4/In_2S_3$ heterostructure for enhanced sonophotocatalytic degradation of pesticides *Ultrason. Sonochem.* **100** 106615
[141] Fatima T, Husain S and Khanuja M 2022 Superior photocatalytic and electrochemical activity of novel WS_2/PANI nanocomposite for the degradation and detection of pollutants: antibiotic, heavy metal ions, and dyes *Chem. Eng. J. Adv.* **12** 100373
[142] Meng J, He M, Li F, Li T, Huang Z and Cao W 2023 Combining Ce-metal–organic framework with CdS for efficient photocatalytic removals of heavy metal ion and organic pollutant under visible and solar lights *Inorganica Chim. Acta* **557** 121701
[143] Wood D, Shaw S, Cawte T, Shanen E and Van Heyst B 2020 An overview of photocatalyst immobilization methods for air pollution remediation *Chem. Eng. J.* **391** 123490
[144] Wang A, Teng Y, Hu X, Wu L, Huang Y, Luo Y and Christie P 2016 Diphenylarsinic acid contaminated soil remediation by titanium dioxide (P_25) photocatalysis: degradation pathway, optimization of operating parameters and effects of soil properties *Sci. Total Environ.* **541** 348–55
[145] Srivastava A and Bandhu S 2022 Biotechnological advancements and challenges in textile effluents management for a sustainable bioeconomy: Indian case studies *Case Stud. Chem. Environ. Eng.* **5** 100186
[146] Jo W-K and Tayade R J 2014 New generation energy-efficient light source for photocatalysis: LEDs for environmental applications *Ind. Eng. Chem. Res.* **53** 2073–84
[147] Kallawar G A, Barai D P and Bhanvase B A 2021 Bismuth titanate based photocatalysts for degradation of persistent organic compounds in wastewater: a comprehensive review on synthesis methods, performance as photocatalyst and challenges *J. Clean. Prod.* **318** 128563
[148] Giram D, Das A B B and Bhanvase B 2023 Comparative study of ZnO–TiO_2 nanocomposites synthesized by ultrasound and conventional methods for the degradation of methylene blue dye *Indian J. Chem. Technol.* **30** 693–704
[149] Bhanvase B A, Shende T P and Sonawane S H 2017 A review on graphene–TiO_2 and doped graphene–TiO 2 nanocomposite photocatalyst for water and wastewater treatment *Environ. Technol. Rev.* **6** 1–14
[150] Mohanta D, Barman K, Jasimuddin S and Ahmaruzzaman M 2021 Encapsulating band gap engineered $CoSnO_3$ mixed metal oxide nanocomposite in rGO matrix: a novel catalyst towards LED light induced photoelectrocatalytic water oxidation at neutral pH *J. Electroanal. Chem.* **880** 114830
[151] Anjum M, Miandad R, Waqas M, Gehany F and Barakat M A 2019 Remediation of wastewater using various nano-materials *Arab. J. Chem.* **12** 4897–919
[152] Alalm M G, Djellabi R, Meroni D, Pirola C, Bianchi C L and Boffito D C 2021 Toward scaling-up photocatalytic process for multiphase environmental applications *Catalysts* **11** 562
[153] Haldar S S, Halder D, Patel G, Singhania A K, R R and Pandey A 2023 Waste fish scale for the preparation of bio-nanocomposite film with novel properties *Environ. Technol. Innov.* **32** 103386

[154] El-Sheikhy R, Al-Khuraif A and Al-Shamrani M 2023 Converting polymer-wastes to green nanocomposites in Saudi Arabia: investigation of fracture parameters (KIc, KIIc, θc, σc) for quality and sustainability evaluation *Vacuum* **217** 112513
[155] Fatimah I, Yanti I, Wijayanti H K, Ramanda G D, Sagadevan S, Tamyiz M and Doong R 2023 One-pot synthesis of $Fe_3O_4/NiFe_2O_4$ nanocomposite from iron rust waste as reusable catalyst for methyl violet oxidation *Case Stud. Chem. Environ. Eng.* **8** 100369
[156] Suryawanshi P L, Gumfekar S P, Kumar P R, Kale B B and Sonawane S H 2016 Synthesis of ultra-small platinum nanoparticles in a continuous flow microreactor *Colloid Interface Sci. Commun.* **13** 6–9
[157] Malkapuram S T, Seepana M M, Sonawane S H, Lakhera S K and Randviir E 2024 ZIF-8 decorated cellulose acetate mixed matrix membrane: an efficient approach for textile effluent treatment *Chemosphere* **349** 140836
[158] Vasseghian Y, Khataee A, Dragoi E-N, Moradi M, Nabavifard S, Oliveri Conti G and Mousavi Khaneghah A 2020 Pollutants degradation and power generation by photocatalytic fuel cells: a comprehensive review *Arab. J. Chem.* **13** 8458–80

IOP Publishing

Photocatalysts for Energy and Environmental Sustainability

Vijay B Pawade and Bharat A Bhanvase

Chapter 5

Photocatalysts for hydrogen evolution

Somnath C Dhawale and Bhaskar R Sathe

The most significant green and sustainable energy resource that can meet forthcoming energy needs is hydrogen (H_2). This is because it has the potential to reduce our dependency on fossil fuels while also serving as the ultimate antidote to environmental degradation. The most promising option is the generation of solar-induced H_2 through water splitting, which uses economical solar energy and has the potential for enhanced solar-to-hydrogen conversion efficiency. In this chapter, the most recent advances in metal-free photocatalysts for H_2 production from water are discussed. It has been established that some materials, including carbon nitride and sulfonate, possess exceptional reactivity towards visible light, a sufficient number of active sites, and appropriate H^+/H_2 reduction potential to operate as effective photocathodes. Furthermore, in order address the issue of soluble mediators and simultaneously optimize the reduction and oxidation processes, the application of a soluble redox mediator in Z-scheme water splitting systems has been studied. Solar fuel is also produced using a number of photophysical and electrocatalytic techniques.

5.1 Introduction

Hydrogen (H_2) is a promising and versatile energy carrier which plays an important role in sustainable and environmentally friendly energy systems [1]. For the production of ammonia and in the petrochemical sector, hydrogen is a crucial and widely utilized chemical raw ingredient. As a result, there is a sharp rise in the annual demand for hydrogen. The three principal industrial processes for the production of H_2 are steam reforming, the gasification of coal, and the electrolysis of H_2O. However, these processes produce H_2 through the use of fossil fuels and therefore release CO_2 as a byproduct as part of these processes. Interestingly, water splitting, especially when powered by renewable energy sources, is often regarded as a more sustainable and environmentally friendly method for H_2 production.

doi:10.1088/978-0-7503-5697-8ch5

This creates an opening for sustainable hydrogen generation, since its primary raw material, water, is plentiful and regenerative. In addition, since the products are hydrogen and oxygen, secondary environmental contamination is avoided. Renewable energy from hydropower, the wind, and the sun can be utilized for the generation of electricity needed for water splitting [2–4] and therefore water splitting has become more significant.

However, only 3%–5% of commercial hydrogen (H_2) is currently produced by water splitting reactions. Water splitting is primarily a kinetically controlled electrochemical process that has an energy-intensive nature since it results from delayed electron transfer at an electrified interface corresponding to two half cells. These correspond to water oxidation and reduction, which are also known as the oxygen evolution reaction (OER) and the hydrogen evolution reaction (HER), respectively [5]. Unfortunately, these processes have exceptionally high energy barriers, which further hampers energy conversion. Unquestionably, the fabrication/synthesis of effective and stable catalysts for the HER and the OER is vital for advancing water splitting technologies [6]. Both electrocatalytic and photocatalytic approaches are being explored to improve the performance of water splitting systems.

Photocatalysts and electrocatalysts for the HER have been thoroughly studied and reported from a number of perspectives during the last few decades [7, 8]. Nowadays, catalysts prepared using noble metals such as Pt and Pd are known to have exceptional catalytic activity, but the cost and scarcity of noble metals can be limiting factors for their commercial application in water splitting technologies. To address this challenge, significant research efforts are being directed toward developing non-precious metal-based systems as alternatives for HER catalysts. The goal is to find materials that are more abundant and cost-effective yet still offer high catalytic activity; examples include transition-metal sulfides, nitrides, selenides, and phosphides [3, 9]. However, the majority of these experience unpredictable agglomeration/dissolution throughout the HER process as well as undesired and/or unstable alterations in their morphology/crystalline structure. Due to the leakage of heavy metals into electrolytes, this low stability may potentially have negative effects on the environment.

The United Nations (UN) claims that energy-related activities are a major contributor to climate change [10]. As a result, research into new renewable energy sources has been included as target 7 of the UN's 17 Sustainable Development Goals. In this regard, a growing focus of basic and applied research has been H_2 generation as a sustainable energy source. In general, hydrogen (H_2) may be produced via various chemical processes as shown in figure 5.1. In comparison, photocatalytic water splitting is reported to be the simplest way. It requires semiconducting particles to create hydrogen and oxygen when they are in solution and exposed to sunlight. It also exhibits thermodynamic feasibility, as shown by the following equations (5.1) and (5.2).

$$H_2O \rightarrow H_2 + \frac{1}{2}O_2 \quad \Delta G^\circ = 238\ \mathrm{mol}^{-1} \tag{5.1}$$

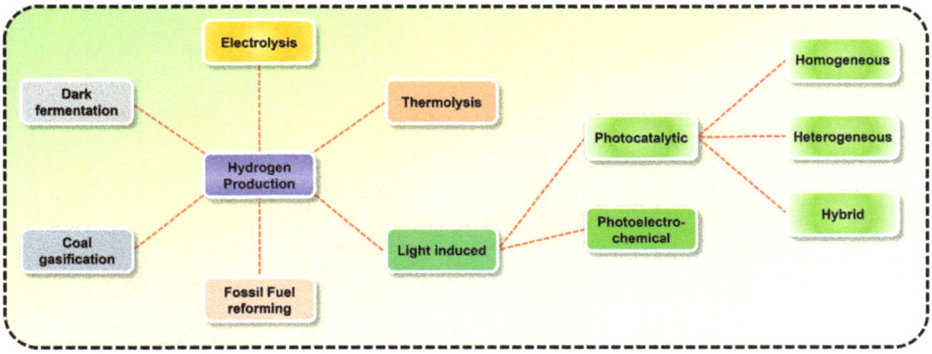

Figure 5.1. A schematic survey of the major feasible H_2-generating routes. The most promising option is solar-induced H_2 generation by water splitting, considered to be a promising and environmentally friendly approach for sustainable energy production.

$$PC\ h\nu > Eg \rightarrow PC^{*} + h^{+} + e^{-} \tag{5.2}$$

$$H_2O + 2h + \rightarrow \frac{1}{2} O_2 + 2H + (+0.82\ \text{V vs normal hydrogen electrode},\ pH = 7) \tag{5.3}$$

$$2H^{+} + 2e^{-} \rightarrow H_2(-0.41\text{V vs normal hydrogen electrode},\ pH = 7) \tag{5.4}$$

When water is split by means of a photocatalyst, the valence band (VB) must have a higher positive potential than the water oxidation potential required for the OER (equation (5.3)). However, in order to drive the reduction processes, a conduction band (CB) potential less negative than the normal H_2 evolution (vs. NHE) potential is needed (equation (5.4)). Ultimately, this procedure offers a pure and sustainable supply of hydrogen without generating toxic gases or other byproducts. The photocatalyst must meet energy gap requirements which are necessary to produce H_2 and O_2 simultaneously; these are considered to be challenging.

Consequently, combining catalysts to form cocatalysts is a tactic frequently used to pursue this goal. Different catalysts have been created for this purpose, including homogeneous, heterogeneous, and hybrid ones. Heterogeneous photocatalysts are well known for a variety of properties, including high chemical stability in a liquid medium, which offers resistance to degradation or corrosion during the chemical reaction. This is important for ensuring the longevity and effectiveness of the catalyst under reaction conditions, reducing operating costs, and enabling the production of specific products.

Recently, a lot of interest in semiconductor metal sulfides has been observed because of their outstanding physical and chemical characteristics. Numerous binary metal sulfides (MS), as well as their derivatives and heterostructures, are now extensively used for a variety of applications [11–16]. The most popular metals used to make sulfide compounds for photocatalytic H_2 production are listed in figure 5.2. The majority of these compounds exhibit extraordinary activity in visible light, enough reactive centers, and the right reduction potential for H_2 production to

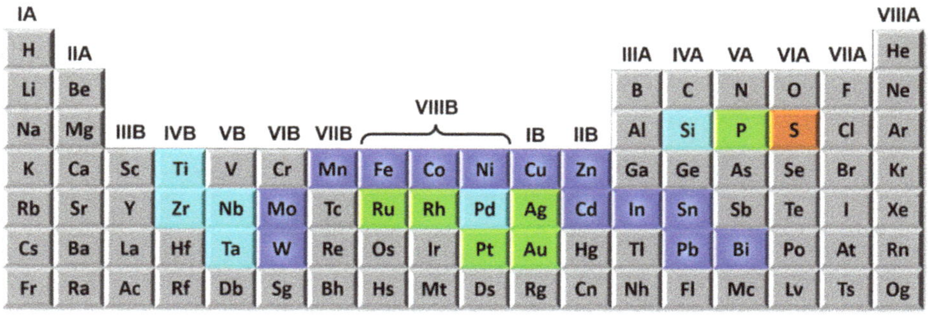

Figure 5.2. A representation of elements in the periodic table whose sulfide compounds are often employed in photocatalytic H_2 production (blue), used for dopants (green), or whose oxides or other compounds (cyan) are used to build heterostructures with MS for H_2 evolution.

function as efficient photocatalysts. In addition, new quantum size effects allow for additional tuning to obtain quicker charge transfer, prolonged excited-state durations, etc [17, 18]. Over the last ten years, a lot of work has been done on boosting solar hydrogen generation using MS semiconductor photocatalysts [19–21]. $CuInS_2$ and $CuGaS_2$ have been demonstrated to have good selectivity and a high photocatalytic H_2 generation rate compared to those of other MS materials [22, 23], particularly when combined with titanium dioxide (TiO_2). The focus of current research is on MS with reduced dimensions. Moreover, research into high H_2 generation is expanding due to the functionalization of two-dimensional (2D) MS structures [24]. One popular method of increasing photocatalytic activity is to utilize metals as dopants or other substances as cocatalysts, as shown in figure 5.2.

The evolution and further advancement of civilization depend on our capacity to harness energy beyond that provided by humans and other animals and on our ability to locate, capture, and utilize energy more and more skillfully. Among the various resources of energy, oil, gas, and coal account for more than 85% of global energy production [1]. A sustainable source equal to 1066 barrels of oil, 108 000 m^3 of natural gas, and 250 tons of coal/second would have to be generated to fulfill the world average energy consumption rate of 17.2 TW [25]. The world's energy consumption is anticipated to quadruple by 2050 due to rising population growth (1.12% per year) and industrialization [26]. An 'energy crisis' will emerge from the quick depletion of finite fossil fuel supplies, even if the present 2.1% annual rate of global primary energy use does not change. In addition, all living things and plants are harmed by the poisonous effluent emissions that result from burning fossil fuels, which also pose a hazard to humanity [27]. Climate change (according to the United Nations Framework Convention on Climate Change (UNFCCC)) poses an unstoppable threat to society. As a result, in addition to the usual choices of afforestation and bioenergy with carbon capture and storage (BECCS), the Paris Agreement asks for the growth of other net CO_2 removal technologies.

The hunt for alternative energy sources is sparked by how we respond to this situation. In order to reduce (or avoid) the anthropogenic effects produced by the combustion of fossil fuels and to address the safety of supply in the short term, it is generally acknowledged that a transformation from nonrenewable fossil fuels to renewable fuel resources is necessary. Solar energy is thought to be a feasible option for producing hydrogen from water, assuming that catalysts with the necessary activity, selectivity, and durability can be created [28]. A noteworthy aspect of H_2 fuel is that it burns cleanly with no production of CO_2 or NO_x. In addition, it can be kept in a storage unit and utilized as needed [29].

In recent years, the necessity to preserve natural resources and limit energy use has prompted the creation of energy storage devices that are renewable and sustainable. Hybrid vehicles and portable electronic gadgets have made use of unconventional energy sources such as batteries, supercapacitors, and fuel cells, which use electrochemical reactions to convert chemical energy into electrical energy [30]. However, H_2 is a regenerative and ecologically friendly energy carrier that has attracted the interest of scientists for the last three decades [31, 32]. Nowadays, H_2 is mostly generated from fossil fuels, primarily through a process known as the steam methane reforming technique, which is both ecologically and economically unfavorable [33]. H_2 is expected to be generated using renewable energy sources; a facile pathway for this purpose is photocatalytic H_2 production based on water splitting driven by solar energy technologies. This leverages solar energy for both energy storage and green H_2 production, which have garnered significant attention due to their potential to address key challenges in sustainable energy. Solar energy conversion is a sustainable solution that prevents global warming and the energy crisis [34]. Following Fujishima and Honda's development of photoelectrochemical water splitting in 1972, several research reports describing the application of various photocatalysts have been published [35]. The application of powder photocatalysts is indeed common in research related to photocatalysis, including water splitting, and environmental remediation [36–41]. Photocatalytic H_2 production is a clean technique that uses semiconductor nanoparticles and abundant sunshine irradiation to manufacture environmentally acceptable fuel. The critical challenge in the photocatalytic H_2 production process is finding highly efficient and stable photocatalysts which are active under visible light and can facilitate both the HER and the OER under visible-light exposure [42]. Moreover, overcoming issues related to quantum efficiency, photostability, and visible-light absorption is critical for advancing the field and making these technologies more viable for real-world applications. In recent decades, the most widely researched photocatalysts for water splitting reactions have been TiO_2, g-C_3N_4, and CdS [34]. Some promising visible-light-driven photocatalysts such as Ta_3N_5, [43, 44], TaON [45], GaN:ZnO [13, 46], sulfides [47–49], CoO [50], and also g-C_3N_4 [51–53] have been discovered and used for H_2 production. However, the solar-to-fuel conversion efficiency is variable and still very low, which hampers practical applications.

Undoubtedly, a review of metal-free photocatalysts would contribute significantly to the development of novel materials for further energy conversion and storage applications [54–56]. Thus, the development of the same would offer a wealth of information to researchers and technologists, allowing them to acquire knowledge about the current state of the subject and to understand the problems that need to be solved in order to forge a future course. However, no review has yet been made available that has addressed a substantial collection of metal-free photocatalysts. A study of this nature is therefore highly appropriate, given the speed of scientific advancement. We anticipate that a comparison of several metal-free photocatalysts would provide information that would allow us to choose, examine, and screen the optimal metal-free photocatalyst combination for high solar-to-hydrogen conversion efficiency.

5.2 The fundamentals of photocatalytic water splitting

According to thermodynamics, the entire water-splitting reaction is an endothermic, energetically uphill reaction with a Gibbs free energy change ($\Delta G°$) of 237 kJ mol^{-1} showed in equation (5.1). Three crucial stages are involved in a photocatalytic water splitting reaction, as follows: (i) the process of producing electron–hole pairs through the photoexcitation of photocatalysts, (ii) the separation and transport of the generated electrons and holes to the surface of the photocatalyst, and (iii) the capture of the generated electrons and holes by active sites on the cocatalysts for consumption in electrode reactions for H_2O splitting reactions, as shown in figure 5.3 [37, 57, 58]. The total efficiency of solar energy conversion is calculated by equation (5.5):

$$\eta = \eta 1 \ \cdot (\eta 2 \ \times \ \eta 3), \tag{5.5}$$

where η_1 is the efficiency of light harvesting, η_2 is the efficiency of charge separation, and η_3 is the efficiency of the surface catalytic reaction.

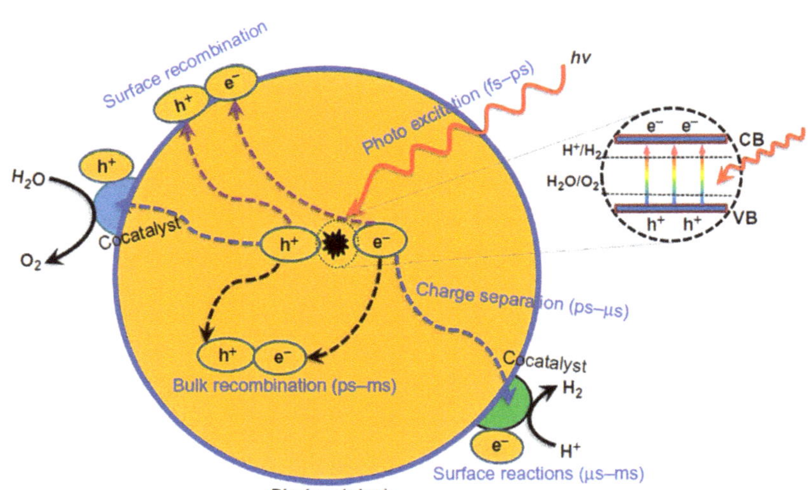

Figure 5.3. The mechanism of semiconductor-based photocatalytic water splitting. Adapted from [57], with permission from the American Chemical Society, and [59], with permission from Elsevier.

These processes typically include light absorption, charge carrier generation, migration, surface reactions, and the production of H_2 and O_2. The claimed water splitting efficiencies are often poor because recombination results in the bulk of the charges being lost.

Certain band structures are found in semiconductors, and their bandgaps are often found to lie in the energy range of 0.2–4.0 eV. If the photon energy of incident light is equivalent to or higher than the bandgap of a semiconductor material, electrons in the VB can be excited to the CB. After photoexcitation, electrons–hole pairs are released and come to the surface for the transfer processes. The holes oxidize H_2O molecules to produce protons and O_2, while the electrons convert the liberated protons to H_2. The surface reduction and oxidation processes that generate H_2 and O_2 resemble those of water electrolysis. The highest occupied molecular orbital (HOMO) and lowest unoccupied molecular orbital (LUMO) energy levels of an organic complex photocatalyst are crucial for driving redox reactions during photocatalysis. Efficient and effective photoinduced electron transfer processes can only be achieved by aligning these energy levels with the redox potentials of the involved species, such as O_2/H_2O, using the correct alignment. Hence, a photon in the optical range of 1000 nm requires a minimum Gibbs free energy of 1.23 eV for water splitting. The electrode interfacial reactions of water splitting, especially for the OER, require the photocatalysts to have bandgaps of more than than 1.23 eV due to the high overpotential.

To convert solar energy efficiently, a semiconductor-based photocatalyst must first have a bandgap suitable for both efficient light absorption and the CB and VB potentials needed for the redox reactions that split water. Second, the ideal photocatalyst should be able to separate and transmit photogenerated charges with remarkable efficiency. Charge separation and recombination are competing processes in photocatalysis. The recombination of photoexcited charges frequently occurs within a short time interval (picoseconds to microseconds). Consequently, the water splitting redox reactions on the surface must take place during the lifetimes of the photoexcited charges [41]. The efficient separation and transfer of photogenerated charges is necessary for photocatalytic water splitting. Any methods that help with charge separation should be helpful in the design and construction of extremely effective photocatalyst systems. Third, the surface-active sites should take up the electrons and holes fast, so they can join the reactions with the adsorbed species. In general, the surface response occurs over a longer period (microseconds to seconds) than the charge formation and separation processes. Therefore, methods for increasing the surface catalytic reactions are also necessary to build photocatalyst systems that work well. An efficient method for hastening the catalytic process and raising the total water splitting efficiency is to load suitable cocatalysts onto the surface of the photocatalyst to create additional active sites. It is important to note that surface catalytic reactions and charge separation are not separate processes but rather have a synergistic correlation. Effective charge separation benefits from quicker surface reactivity and vice versa.

5.3 Quantum efficiency and solar-to-H_2 efficiency

In photochemistry, the quantum yield is a measure of the efficiency of a particular process, indicating the rate at which a certain event occurs for each photon absorbed in a given unit of time. The quantum yields of light emission, product creation, reactant disappearance, and other photochemical and photophysical processes that occur in photochemical reactions are frequently measured by photochemists [60]. Comparing the activities of photocatalysts is often challenging due to the substantial dependence of their photoactivity on experimental circumstances, such as light intensity and reaction temperature [38]. It is important to note that dividing the reacted electrons by the absorbed photons yields the true quantum efficiency (the internal quantum efficiency (IQE)). However, due to light scattering and loss, determining the quantity of photons absorbed by a particular photocatalytic system is difficult. In general, when H_2 or O_2 gas is produced by a certain photocatalyst system, the number of incident photons can be measured using a thermopile. Therefore, the gain in terms of quantum efficiency is the apparent quantum efficiency (AQE) at a particular monochromatic wavelength, which may be calculated using the following equation for a given photocatalyst [61].

$$\begin{aligned} \text{AQE} &= \frac{\text{Number of reactive electrons}}{\text{Number of incident photons}} \times 100\% \\ &= \frac{2 \times \text{ Number of evolved } H_2 \text{ molecules}}{\text{Number of incident photons}} \times 100\% \\ &= \frac{4 \times \text{ Number of evolved } O_2 \text{ molecules}}{\text{Number of incident photons}} \times 100\% \end{aligned} \tag{5.6}$$

$$\begin{aligned} \text{IQE} &= \frac{\text{Number of reactive electrons}}{\text{Number of absorbed photons}} \times 100\% \\ &= \frac{2 \times \text{ Number of evolved } H_2 \text{ molecules}}{\text{Number of absorbed photons}} \times 100\% \\ &= \frac{4 \times \text{ Number of evolved } O_2 \text{ molecules}}{\text{Number of absorbed photons}} \times 100\% \end{aligned} \tag{5.7}$$

A realistic metric for evaluating photocatalyst performance is the solar-to-hydrogen (STH) efficiency, which may be computed using the formula below:

$$\text{STH} = \frac{\text{Output energy as } H_2 \text{ gas}}{\text{Energy of incident solar light}} = \frac{r_{H_2} \times \Delta G}{P_{sun} \times S} \times 100\% \tag{5.8}$$

where R_{H_2} is the rate of H_2 production (mmol s^{-1}), P_{sun} is the solar energy flux (mW cm^{-2}), S is the reactor area (cm^2), and G is the Gibbs free energy (kJ mol^{-1}).

5.4 Overall water splitting (hydrogen evolution/oxygen evolution)

The total water splitting efficiency can be used to quantify the photocatalytic water splitting capacity in the absence of sacrificial chemicals. H_2 and O_2 are produced in a 2:1 stoichiometric ratio during the entire process (equations (5.9) and (5.10)). Sacrificial species are added to the relevant half-cell reaction to quickly control the photogenerated holes and/or electrons, which is the rate-limiting phase (figure 5.4). For instance, in the HER (equations (5.11) and (5.12)), lactic acid, methanol, or triethanolamine is commonly employed as a hole scavenger and in the OER (equations (5.13) and (5.14)), Fe^{3+}, IO_3, or Ag^+ is used. Two considerations must be addressed for half-reactions: (1) determining the thermodynamic suitability of a photocatalyst's valence or conduction bands for proton reduction or water oxidation, and (2) calculating real-world proton reduction and water oxidation kinetic reaction rates.

The overall water splitting reaction:

$$\text{Reduction: } 4H^+ + 4e^- \rightarrow 2H_2 \tag{5.9}$$

$$\text{Oxidation: } 2H_2O + 4h^+ \rightarrow 4H^+ + O_2 \tag{5.10}$$

The H_2-generating half-reaction in the presence of CH_3OH:

$$\text{Reduction: } 6H^+ + 6e^- \rightarrow 3H_2 \tag{5.11}$$

$$\text{Oxidation: } H_2O + CH_3OH + 6h^+ \rightarrow CO_2 + 6H^+ \tag{5.12}$$

The O_2-generating half-reaction in the presence of Ag^+:

$$\text{Reduction: } 4Ag^+ + 4\,e^- \rightarrow 4Ag \tag{5.13}$$

$$\text{Oxidation: } 2H_2O + 4h^+ \rightarrow 4H^+ + O_2 \tag{5.14}$$

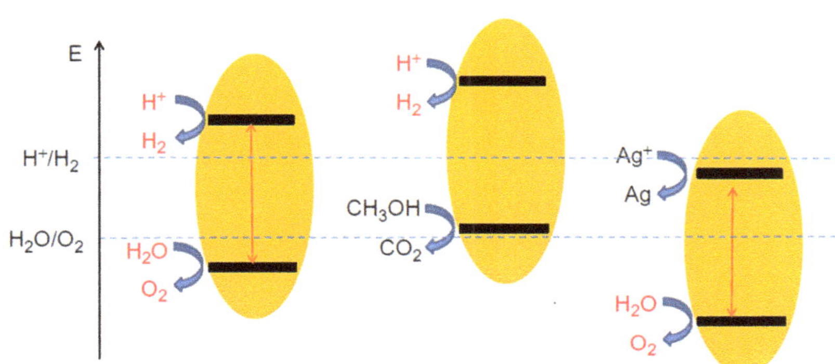

Figure 5.4. A schematic showing the half-reactions (the OER and HER) of photocatalytic water splitting in the presence of sacrificial agents. Reprinted from [59], Copyright (2017), with permission from Elsevier.

5.5 The basic concepts of photocatalysis

Thermal catalysis and photocatalysis are two subfields within the larger field of catalysis. The International Union of Pure and Applied Chemistry (IUPAC) defines photocatalysis as 'a change in the rate of a chemical reaction or its initiation in the presence of a substance—the photocatalyst—that absorbs light and is involved in the chemical transformation of the reaction partners.' At the turn of the 20th century, researchers believed that light irradiation might be utilized to stimulate a process. Between 1900 and 1920, Giacomo Ciamician was the first scientist to try to understand how light impacts chemical reactions [62]. However, the development of the field of photocatalysis began primarily in the 1970s [35, 63].

Figure 5.5(a) shows the catalytic conversion of a reactant (R) to a product (P) via active intermediate species denoted by I′ and C′. In the case of thermal catalysis, the activation of R occurs through the physical process of energy transfer, whereas in photocatalysis, it occurs when the excited state (C*) of a photocatalytic system transfers an electron, as schematically shown in figures 5.5(b) and (c). In photocatalysis, the electrical activation of R results in the production of an intermediate I (a radical or radical ion), which is theoretically also capable of being created by a thermal process.

After the electrons are transferred, C* deactivates to form C°. The intermediate I regenerates C in the final state and provides the end product P (potentially through

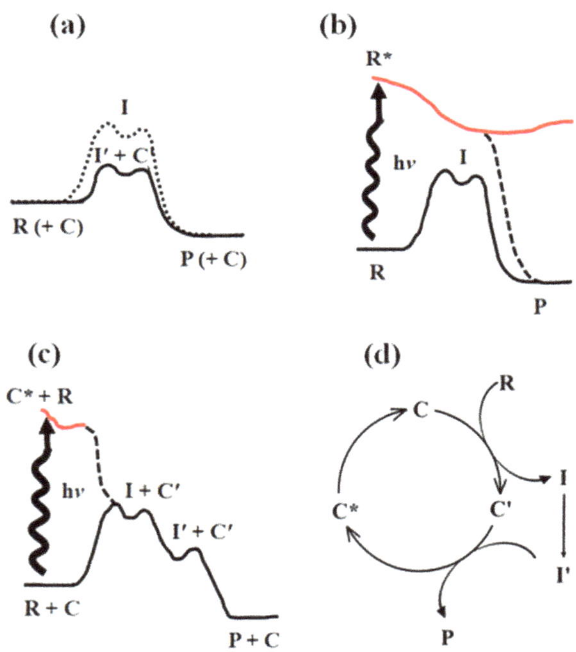

Figure 5.5. (a) A thermal reaction (R→P) that C catalyzes through the intermediate I°; (b) a photochemical reaction in which the chemical reaction starts when the reagent R* is in the excited state; (c) a photocatalyzed reaction in which R undergoes all of its chemical transformations on the ground-state surface, but the catalyst C is active only in the excited state; and (d) C serves as a photocatalyst in the R→P reaction. Reproduced from [64] with permission from the Royal Society of Chemistry.

further intermediates I′). In any reaction pathway, although the chemical route partially occurs on surface R* as depicted in figure 5.5(b), the whole chemical transformation of R to P takes place on the surface with the lowest potential energy. The process described, in which a new reaction pathway becomes available on the ground-state surface because of the absorption of light, is commonly referred to as photocatalysis [64, 65].

The shift in Gibbs energy (ΔG) is another significant characteristic that sets photocatalytic reactions apart from conventional thermal catalytic processes. ΔG may be neutral or favorable. Energy release is implied by a negative ΔG, whereas energy absorption or storage is denoted by a positive ΔG. Normal catalysts can only be used to catalyze negative ΔG reactions. This is due to the fact that a heated catalyst accelerates a chemical reaction that would continue spontaneously with a negative Gibbs energy change by minimizing the activation barrier of the transformation and altering the intermediate states. However, in photocatalytic processes, both scenarios are feasible [66]. For instance, even when ΔG is positive, an overall redox reaction can still happen, which explains why a reaction with positive ΔG cannot proceed spontaneously. Simply put, the reduction and oxidation processes must be physically or chemically distinct from one another for the reaction between the two products to occur without producing any net products. Under these conditions, the reactions of e^- and h^+ with the oxidant (ΔG_e) and the reductant (ΔG_h) must have a negative Gibbs energy change; that is, the reactions must occur spontaneously and be followed by photoexcitation, as shown in figure 5.6. In order to have a negative change in Gibbs energy for both processes, the photocatalyst material's CB bottom must be more cathodic and its VB top should be more anodic than the reference system potentials of both the oxidant and reductant species. Moreover, for both reactions, i.e. those with positive ΔG and those with negative ΔG, the reduction and oxidation must be separately performed by electrons (e^-) and holes (h^+), respectively [66].

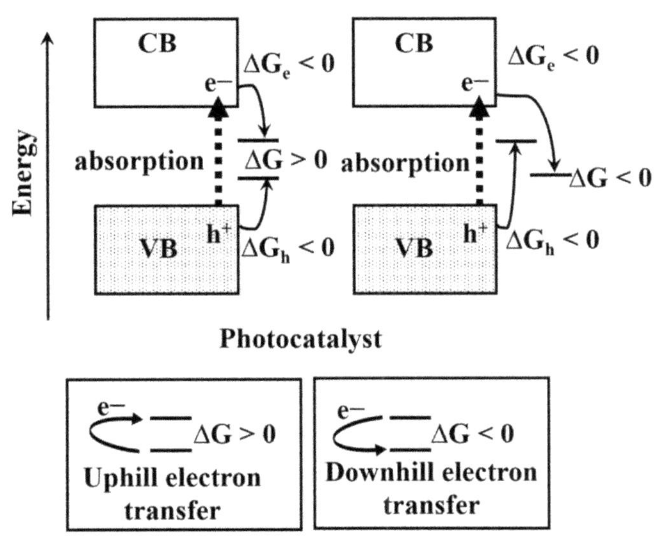

Figure 5.6. The change in Gibbs energy during photocatalytic reactions. Adopted from [66], Copyright (2010), with permission from Elsevier.

5.6 Elements of photocatalytic hydrogen generation physics and chemistry

Figure 5.7 illustrates the various photophysical and electrochemical processes that are involved in the production of photocatalytic hydrogen. The photophysical process includes the production of charge carriers and photon absorption as well as their separation and transport to the reaction sites, whereas the electrochemical process also includes redox reactions for water splitting. A photocatalyst absorbs photons of specific wavelengths when exposed to light, based on its inherent bandgap. Semiconductor photocatalysts can absorb photons that strike them at energies equal to or greater than their bandgap energy. An electron in the valence band gains energy from the absorbed photon, which enables it to cross the bandgap and move into the conduction band. A positively charged hole is left in the valence band as a result of this action. As a result, an electron–hole pair is produced, with the hole now in the valence band and the photoexcited electron now in the conduction band. Various photoinduced processes are made possible by the fact that the photoexcited electron and hole can move independently within the semiconductor material. Photon absorption results in the production of excited holes and electrons. To maintain equilibrium, excited electrons in the CB may radiatively or non-radiatively return to

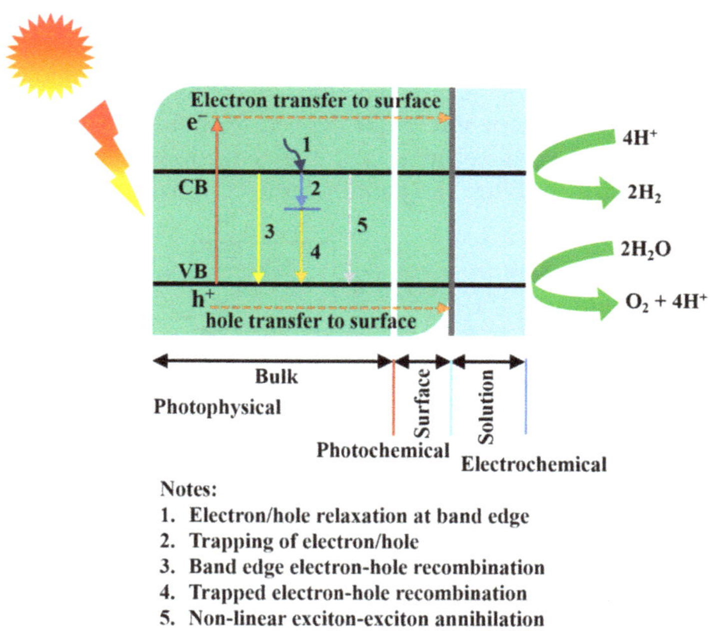

Figure 5.7. Photocatalytic water splitting involves three main dynamical processes (photophysical, photochemical, and electrochemical). The bandgap is excited by the following: (1) electron/hole relaxation at the band edges; (2) electron/hole trapping due to defects or surface states; (3) band-edge electron–hole or exciton recombination; (4) trapped electron–hole or relaxed exciton recombination; and (5) nonlinear exciton–exciton annihilation for electron/hole relaxation. Reprinted with permission from [71]. Copyright (2019) American Chemical Society.

the VB and recombine with holes. During recombination, the open electron–hole pairs are eliminated. This is an inevitable photophysical process, regardless of the type of semiconductor used as a photocatalyst [67, 68]. Recombination is an important cause of low QE in photocatalysts [69, 70].

The remaining electrons and holes must subsequently be delivered to the surface, where the HER and OER are carried out. Photoexcited carriers can be transported to the active surface sites via diffusion, or electric fields can be applied to the interface between the semiconductor and the electrolyte or between the semiconductor and the cocatalyst [70]. Excitons are produced on a timescale of less than 100 fs. Excitons generally only have lifespans of a few hundred picoseconds on average. Electrons can diffuse within a semiconductor after they are photoexcited from the VB to the CB. This diffusion involves the movement of electrons through the crystal lattice. Electron diffusion typically occurs on the order of a few picoseconds. A number of variables, including the diffusion coefficient and the distance that electrons must travel before scattering or recombining, can affect the diffusion time. On the other hand, the analogous transition time for holes is often even shorter, typically ranging from 100 to 300 femtoseconds (fs). Holes, as positive carriers, can move relatively quickly in comparison to electrons [72, 73]

5.7 Systems for producing hydrogen via photocatalysis

Water-splitting reactions can result from photocatalysis in two ways: (i) partially, in which only the reduction process is involved, or (ii) completely, in which both reduction and oxidation events are involved at the same time. Triethanolamine, alcohol, and other sacrificial electron donors and hole scavengers can be employed to produce the water splitting half-reaction. These agents help in realizing the reduction or oxidation reactions separately. When sacrificial agents are present, the photogenerated holes in the photocatalyst oxidize the sacrificial agent rather than water. The photocatalyst's electrons become more enriched as a consequence. Because this process depends on sacrificial agents, it is crucial to understand that it does not represent water splitting. The stoichiometric separation of water into H_2 and O_2 without the use of sacrificial agents is known as 'water splitting.' When sacrificial agents are present, the process should be differentiated from genuine water splitting since it entails the oxidation of the sacrificial agent. Figure 5.8 depicts examples of both one-step (single-photon) and two-step (two-photon) systems that are available for generic water splitting under visible-light irradiation [38]. The two-photon systems are also known as Z-scheme processes. For one-step complete water splitting, Z-scheme photocatalyst systems use a single semiconductor with band-edge positions that suit the H^+/H_2 and O_2/H_2O potentials [74]. A single semiconductor is used in Z-scheme systems, and its band-edge positions are chosen so that they align with the potentials for the evolution of both oxygen (O_2/H_2O) and hydrogen (H^+/H_2). This makes it possible to split water completely in one process. Separate hydrogen evolution photocatalysts (HEPs) and oxygen evolution photocatalysts (OEPs) are used in two-step (Z-scheme) water splitting systems to evolve H_2 and O_2, respectively.

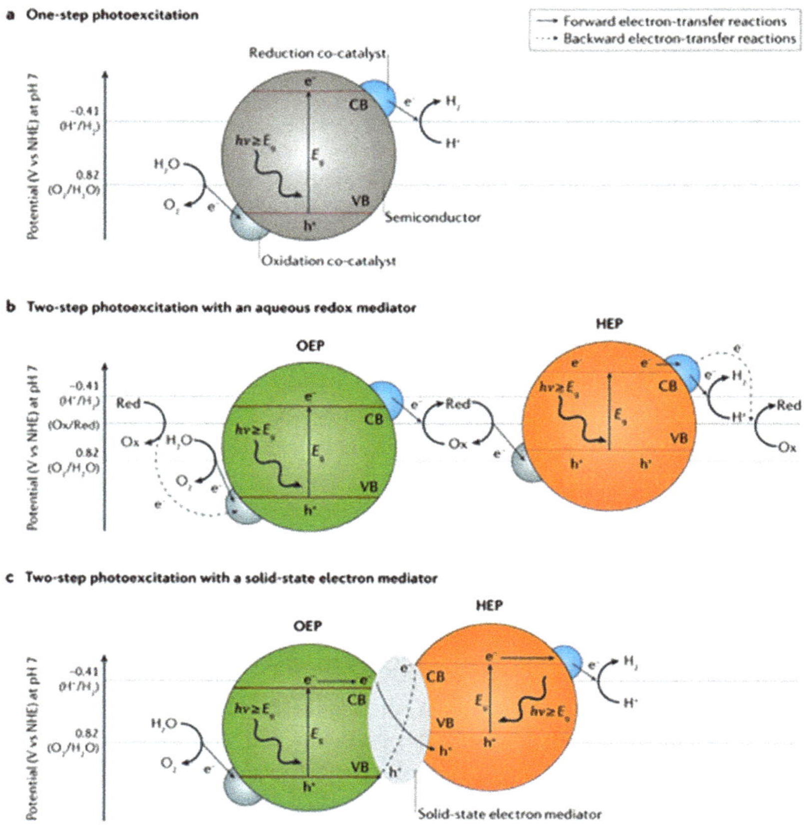

Figure 5.8. A schematic representation of several photocatalytic water splitting systems. Reprinted from [75], Copyright (2017), with permission from Nature.

It is important to note that a Z-scheme containing separate HEP and OEP photocatalysts for total water splitting is not a set rule. Semiconductors with HER or OER activity can be employed in this system. The Z-scheme thereby provides a stronger driving force by lowering the thermodynamic requirements for water splitting. The OEP and HEP in a Z-scheme system continue to have photoexcited electrons and holes, respectively. This is unlike a one-step system in which the entire photocatalytic cycle is completed in a single stage. The recombination of holes in the HEP and electrons in the OEP is necessary in the Z-scheme system to complete the photocatalytic cycle. Figures 5.8(b) and 5.8(c) illustrate the solid-state electron mediator and aqueous redox mediator that are used to facilitate this recombination. The Z-scheme system needs eight electrons to complete the cycle in two stages, as opposed to a one-step system that may require four electrons to complete the photocatalytic cycle. A one-step photoexcitation process produces twice as much H_2 and O_2 as a two-step photoexcitation system at given apparent quantum yield (AQY) and unity light absorption values [75]. Because of its susceptibility to reverse electron transfer, the Z-scheme system presents greater kinetic challenges than the

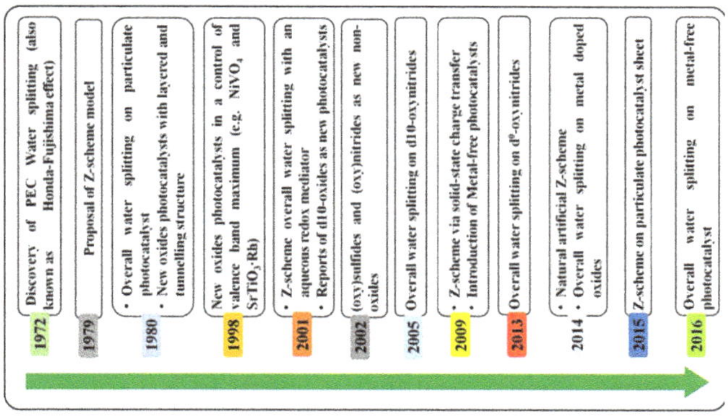

Figure 5.9. A list of significant advancements in photocatalytic water splitting over time.

one-step system. The effective functioning of a Z-scheme water splitting process depends on controlling the interparticle charge transfer between the HEP and the OEP in addition to preventing backward electron transmission.

5.8 Current advances in photocatalytic water splitting

Bard and his colleagues carried out a major push to advance photocatalytic water splitting in the late 1970s [76]. In summary, they discussed likely ways to integrate semiconductor electrodes and heterogeneous photocatalysts in PEC cells, emphasizing both the fundamental principles and practical applications of these components in the context of solar-driven processes. A significant amount of research has been done in the search for extremely reliable, robust, and affordable heterogeneous photocatalysts since 1980 [77]. Wide-bandgap metal oxides with specific cations, such as TiO_2 (titanium dioxide) with Ti^{4+}, are indeed known for their efficient photocatalytic properties under UV light. The key feature of these materials is their wide bandgap, which permits the absorption of UV light and generates electron–hole pairs, initiating photocatalytic reactions. Other metal oxides mentioned in the research literature, such as Zr^{4+}, Zn^{2+}, Mo^{6+}, Ce^{4+}, Nb^{5+}, Ge^{4+}, In^{3+}, Ta^{5+}, W^{6+}, Ga^{3+}, Sn^{4+}, Sb^{5+}, etc. may exhibit similar photocatalytic properties depending on their specific electronic configurations and bandgap characteristics. The efficiency of a photocatalyst is influenced by factors such as the bandgap width, charge carrier mobility, and recombination rate [28, 38]. It is worth noting that while TiO_2 has been widely studied and reported as an active photocatalyst under UV light because its energy levels suffice to launch the water splitting process [78]. Nitrides with d^{10} metal ions are also effective water splitting photocatalysts when exposed to UV light. In photocatalysis, non-oxide powder photocatalysts such as germanium nitride (Ge_3N_4) and cocatalyst-doped gallium nitride (GaN) have garnered interest [13, 79, 80].

With energy gaps between 4 and 3 eV, wide-bandgap photocatalysts are less effective at converting solar energy, since UV light makes up just 4% of the solar spectrum [81]. The photon absorption of wide-bandgap photocatalysts is enhanced by bandgap engineering approaches such as the use of visible-light sensitizers and metal/nonmetal ion doping [82–87]. These methods increase photon absorption, but they also affect the redox potentials, since the elements utilized to narrow the bandgap can also serve as sites of recombination for charge carriers produced by photosynthesis [84, 86]. Noble metals (Au, Ag, Pt, Rh, etc.) are effective sensitizers. Metal nanoparticles, especially Ag and Au, are used as cocatalysts and are gaining popularity as plasmonic photocatalysts [88, 89]. Large-scale uses are also limited by the expense of noble metals. Organic dyes are hindered by their instability; they break down quickly under longer-term irradiation, limiting their practicality [90]. The creation of photocatalysts that are active under visible light is essential for effective water splitting. When a sufficient electron acceptor is present, oxide photocatalysts such as WO_3 can be stable photocatalysts for O_2 evolution under visible light [91, 92].

The CB and VB positions in metal oxide photocatalysts are crucial for H_2O oxidation and reduction, respectively [93]. Strategies for making wide-bandgap metal oxides sensitive to visible light include doping them with transition metals, valence-band control using anions or p-block metal ions, and spectral sensitization [38]. In conclusion, the difficulties are in finding economical substitutes for noble metals, creating stable photocatalysts for successful water splitting that are active under visible light, and finding a balance between bandgap engineering to maximize absorption and minimizing recombination losses. The goal of ongoing research in these fields is to overcome these obstacles and open the door to useful uses for solar energy conversion.

A donor or acceptor level forms in the forbidden gap when a transition-metal cation or anion is doped into a metal oxide that has a wide bandgap; this acts as a focus for the absorption of visible light. Instead of introducing an energy band into the host material, doping introduces a distinct energy level. This may impede the fast movement of photogenerated carriers and thus affect the effectiveness of the photocatalytic process. The reason for maintaining charge balance is that dopants can function as recombination centers [94]. The production of a VB by orbitals other than those related to O2p is crucial for oxide photocatalysts to function in the presence of visible light [95, 96]. This is required to facilitate effective charge separation and allow for the absorption of visible light. Along with doping, wide-bandgap oxide semiconductor photocatalysts have been investigated in combination with visible-light sensitizers to introduce electrons into the conduction band for the formation of H_2. Small-bandgap semiconductors and organic dyes are common sensitizers. Sensitizers need to have a higher negative excited-state potential than conduction band potential of the metal oxide photocatalyst in order to effectively inject charge [97]. This ensures efficient injection of electrons into the CB, facilitating the photocatalytic reaction. In summary, while doping with transition metals introduces the levels required for visible-light absorption, it comes with challenges related to charge migration and recombination. The use of sensitizers is an alternative approach in which careful selection is needed to ensure effective charge injection into the metal oxide photocatalyst's conduction band. These considerations are essential for the fabrication of photocatalysts that are active under visible light for various applications, including the evolution of H_2. Further additions are supported oxide nanostructures for photoactivated H_2 synthesis [98], such as Ag–ZnO nanocomposites [99], Si–Co_3O_4 nanopyramids [100], CuO–ZnO nanorod arrays, etc [101].

Metal sulfides are also appealing photocatalysts for visible light. There have been several reports of metal sulfide photocatalysts [49, 102–104]. The VB of sulfide photocatalysts is made up of S3p orbitals, which are more negative than O2p orbitals. For instance, CdS is a renowned photocatalyst for visible light that has a small bandgap (2.4 eV). It possesses the necessary band levels for water reduction and oxidation in the presence of sacrificial agents [105]. ZnS, which has an energy gap of 3.6 eV, can be doped with different metal cations to exhibit strong photocatalytic activity for H_2 evolution from aqueous solutions that contain S^{2-} and/or $SO_3{}^{2-}$ as electron donors when exposed to visible light [106]. Under visible light, H_2 evolution activity has been demonstrated in solid solutions including $AgGa_{0.9}In_{0.1}S_2$, CdS–ZnS, $AgInS_2$–ZnS, $CuInS_2$–ZnS, and $CuInS^{2-}AgInS_2$–ZnS [49, 104].

Despite this, in the presence of visible light, sulfur photocatalysts show excellent band levels for water splitting. But rather than O_2 evolution, they encounter difficulties with instability and deactivation, which are mainly caused by photocorrosion or self-oxidation [107]. The bandgap of sulfide materials is narrowed by a VB that results from N2p and S3p orbitals in addition to O2p orbitals. These materials have a bandgap energies in the range of 1.7–2.5 eV, which corresponds to an absorption band of 500–750 nm [38]. Unlike (oxy)nitrides and (oxy)sulfides, materials doped with sulfur or nitrogen are referred to as VB-controlled photocatalysts. Instead of distinct impurity levels, the constituent anion component creates a VB. In strong acids such as aqua regia and highly concentrated H_2SO_4, valence-band-controlled photocatalysts, such as those doped with nitrogen or sulfur, are unstable [84]. (Oxy)nitrides may be partially degraded by photogenerated holes during the initial phases of the photocatalytic process, which results in reduced N_2 evolution. Nevertheless, as reaction progresses, N_2 generation becomes fully controlled. While (oxy)nitrides exhibit stable H_2O reduction and oxidation processes, achieving the required total water splitting performance is hindered by low levels of H_2 evolution. Several (oxy)nitrides that have d0 metal cations, such as Ti^{4+}, Nb^{5+}, and Ta^{5+}, are ineffective in H_2 and O_2 evolution when sacrificial agents are absent [108, 109]. In conclusion, problems arise from the instability of sulfide photocatalysts despite their potential band levels for water splitting. Valence-band-controlled photocatalysts, particularly those doped with nitrogen or sulfur, face stability issues in strong acids, and (oxy)nitrides have limitations in achieving total water splitting performance, especially when sacrificial agents are absent. Ongoing research aims to address these challenges and optimize the effectiveness of these materials in photocatalytic applications.

In 2009, Wang *et al* [51] introduced g-C_3N_4 as a metal-free photocatalyst for water splitting that is active under visible light. g-C_3N_4 is characterized by a medium bandgap of 2.7 eV, which allows it to absorb visible light. It exhibits strong physiochemical stability and has a changeable band structure. g-C_3N_4 can be synthesized in a single step by polymerizing inexpensive feedstocks such as cyanamide, urea, thiourea, melamine, and dicyandiamide [110]. The graphitic planes of g-C_3N_4 consist of tri-s-triazine units linked together by planar amino groups.

The response of g-C_3N_4 under visible light is attributed to electron transfer from the VB, filled by N2p orbitals, to the CB, populated by C2p orbitals. This electronic structure allows g-C_3N_4 to harness visible light for photocatalytic processes. The copolymerization of dicyandiamide with barbituric acid is mentioned as a way of extending the absorption band of g-C_3N_4 to 750 nm. Even without a cocatalyst, g-C_3N_4 has demonstrated photocatalytic H_2 evolution from water, particularly at wavelengths greater than 420 nm [108]. This suggests that g-C_3N_4 has inherent photocatalytic activity for water splitting in the presence of visible light. In summary, g-C_3N_4, with its medium bandgap, stable physiochemical properties, and adjustable band structure, has emerged as a notable visible-light-activated photocatalyst for water splitting. Its synthesis from readily available feedstocks and the ability to extend its absorption band further contribute to its potential in photocatalytic applications.

A new class of photocatalysts active under visible light has the potential to be formed from certain elemental semiconductors, such as Si, P, S, and Se, in addition to compound photocatalysts [111]. Among these, crystalline Si (1.12 eV) has attracted substantial research. Silicon may not be considered to be a material of interest for specific applications in which stability in aqueous environments and suitable band-edge positions are critical. Researchers and engineers often explore alternative materials or modify the properties of silicon to overcome these limitations and enhance its performance in various applications [112]. The photocatalytic activity of red P was initially shown by Wang *et al* [113]. They discovered that crystalline monoclinic red P may generate H_2 from water. Recent research suggests that amorphous sulfur may hold promise as a visible-light-activated photocatalyst. Ongoing studies and advancements in materials science may further elucidate the potential applications and optimization of this materials for photocatalysis [114]. However, it has low hydrophilicity, which reduces its photocatalytic activity. This issue might be solved by modifying the sulfur crystal structure. A material is heated to a certain temperature during the annealing process and subsequently cooled, usually to eliminate imperfections, promote crystallization, or improve the material's properties. In the case of amorphous selenium, the transition to the crystalline form at higher annealing temperatures may impact its photocatalytic properties or stability. This information underscores a challenge associated with using amorphous selenium as a photocatalyst and highlights the importance of understanding and controlling the annealing process to maintain the desired amorphous structure for optimal performance [115].

The development of primary photocatalysts is still in its early stages and suffers from numerous challenges. The more intriguing features of elemental semiconductors might be studied if unique methodologies could be developed. Creating heterojunctions of elemental photocatalysts or forming heterostructures with corresponding sulfides, phosphides, and selenides can lead to unexpected photocatalytic activity. This approach involves combining different materials to take advantage of their complementary properties, potentially synergizing their photocatalytic performance. Investigating specific cocatalysts that may work well with elemental photocatalysts has been mentioned as another option for enhancing performance. Cocatalysts can facilitate charge separation and improve the overall efficiency of photocatalytic processes. A recent addition to metal-free photocatalysts involves organic compounds with C–C/C–H bonds in their π-conjugated organic frameworks. These organic compounds, when utilized as photocatalysts, contribute to the expanding range of materials for light-driven reactions [116–119]. Figure 5.9 depicts the chronology of important advancements in photocatalyst research.

5.9 Overall water splitting using photocatalysts

While designing a photocatalytic water splitting system, factors such as light absorption, photogenerated charge separation, and the use of suitable cocatalysts for catalytic processes should be considered. It is easy to categorize photocatalysts capable of overall water splitting into two main methods, as shown in Figure 5.10. One method involves employing a single photocatalyst to split water into H_2 and O_2 without the need for redox mediators. It is true that the two-step excitation process described is similar to the Z-scheme charge separation approach found in natural photosynthesis systems. It involves separate HEPs and OEPs in addition to an electronic redox mediator.

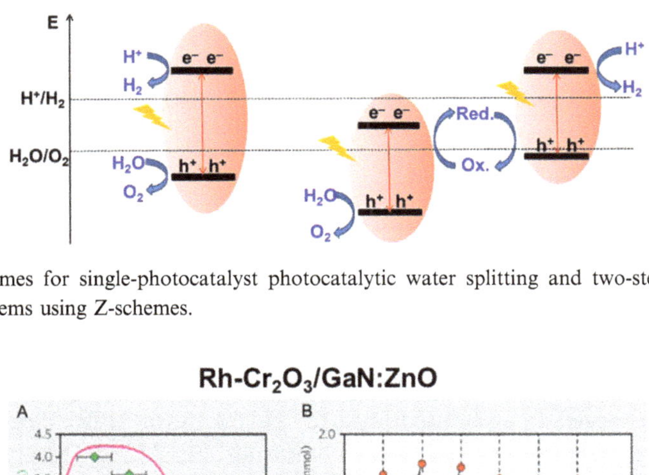

Figure 5.10. Schemes for single-photocatalyst photocatalytic water splitting and two-step photoexcitation photocatalyst systems using Z-schemes.

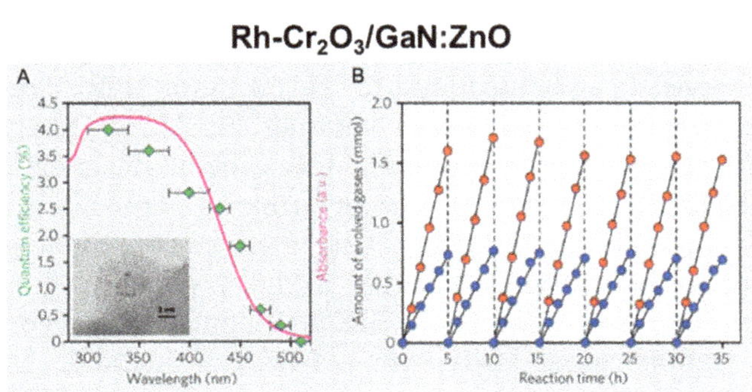

Figure 5.11. Rh–Cr_2O_3/GaN:ZnO—an example of a single photocatalyst for photocatalytic overall water splitting. Adapted from [13], Copyright (2006), with permission from Nature.

5.9.1 Photocatalyst systems with a single photocatalyst

It has been found that there are several single-photocatalyst systems for general water splitting. For instance, a La-doped NiO-$NaTaO_3$ photocatalyst system demonstrated excellent photocatalytic activity for water splitting, with a notable quantum efficiency of 56% at 270 nm [120]. It had a nano-step surface structure for the spatial separation of electrons and holes. Using diluted $CaCl_2$ solution, a Zn-ion-doped Ga_2O_3 photocatalyst was created which exhibited a substantially improved AQE of 71% when illuminated at 254 nm [121]. An Al-doped strontium titanate ($SrTiO_3$) treated with $SrCl_2$ flux has been reported; the AQE of this material was 30% at a wavelength of 360 nm when used for the overall water splitting process [122]. Moreover, gallium nitride (GaN) was loaded with zinc oxide (ZnO) and featured a cocatalyst combination of rhodium (Rh) and chromium oxide (Cr_2O_3). This material was described as responsive to visible light (figure 5.11), suggesting its ability to utilize visible light for the water splitting process. The reported AQE for this GaN:ZnO photocatalyst loaded with Rh/Cr_2O_3 cocatalyst was as high as 5.9%, and the measurement was obtained at a wavelength of 420 nm [13]. Other materials reported include $In_{1x}Ni_xTaO_4$ [123], $LaMg_xTa_{1x}O_{1+3x}N_{23x}$ [124], nitrogen-doped graphene oxide quantum dots [125], and InGaN/GaN nanowire [126]. While progress has been made in total water splitting using photocatalysts, the efficiency of particulate photocatalysts in achieving solar-to-hydrogen conversion remains a significant hurdle. Overcoming this challenge is crucial for realizing the full potential of photocatalytic water splitting as a sustainable and efficient method for hydrogen production.

5.10 Hybrid photosynthesis system: combining natural and artificial approaches

Natural systems offer a paradigm for processing and storing solar energy in the form of chemical fuels such as cellulose and glucose [127]. In recent years, it has also been possible to create hybrid systems by combining some natural photosystem components with synthetic photocatalysts. A two-step photoexcitation natural–artificial hybrid photosystem was built, wherein the $[Fe(CN_6)]^{3-}$ = $[Fe(CN_6)]^{4-}$ donor/acceptor pair shown in figure 5.12 acted as an electronic mediator, a photosystem II (PS-II) membrane made from fresh spinach served as a catalyst for O_2 evolution, and $SrTiO_3$:Rh served as a photocatalyst for H_2 evolution [128]. The hybrid system may effectively perform overall water splitting (at H_2/O_2 levels close to the stoichiometric ratio) when exposed to sunlight. In general, the integration of natural biophoton systems, particularly PS-II, with artificial photocatalysts for efficient charge transfer has promising prospects. This interface could hold significance for the development of biohybrid systems or artificial photosynthesis technologies aimed at sustainable energy conversion. The development of a hybrid photoanode that combines cyanobacterial PS-II with a Fe_2O_3 photoanode has also greatly increased the efficiency of photoelectrochemical water splitting [129]. This is because effective electron transfer from PS-II to Fe_2O_3.

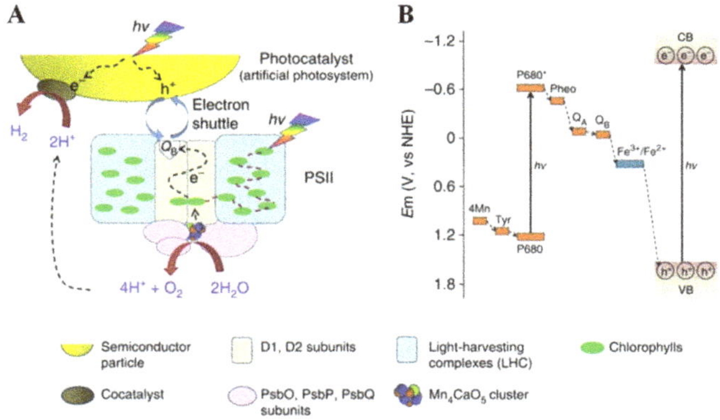

Figure 5.12. PS-II and artificial photocatalysts integrated into two-step photoexcitation natural–artificial hybrid photosystems. Adapted from [128], Copyright (2014), with permission from Nature.

5.11 Conclusions and future perspectives

The most recent developments in metal-free photocatalysts for water-to-hydrogen synthesis were reviewed in this chapter. To make things easier for the reader, we separated the proposed photocatalytic systems into four groups: elemental, binary, ternary, and organic. Among metal-free photocatalysts, carbon nitride is the most well-known and has AQE values of more than 60%, according to recent studies. As a photocatalyst of the future, carbon nitride exhibits excellent potential in terms of scalability, stability, and selectivity. On the other hand, the development of organic and elemental photocatalysts is still in its infancy. For improved quantum efficiency and long-term stable operation, a thorough investigation is needed to enhance their bulk and interfacial electron–hole separation, followed by their transport, and the mediators' redox behavior. Improved photocatalytic performance may be achieved by fine-tuning crystal morphologies and structural features at the nanoscale level, resulting in nanostructured photocatalysts with the desired physicochemical properties. Because of their highly specialized surface areas, easily accessible and interconnected porous networks, efficient charge transfer kinetics, oxidant–reductant interactions, and the active sites of their guest species, nanostructured photocatalysts are an efficient means of collecting light. The combined effects of these mechanisms greatly increase photocatalytic activity. While a lot of work has gone into studying different carbon nitride nanostructures, further investigation is needed to determine the relationship between structure and property and how it affects catalytic performance in other kinds of metal-free photocatalysts. Moreover, the interest in understanding the photophysics and surface processes of organic photocatalysts in aqueous solutions reflects a pursuit of knowledge critical for advancing the field of photocatalysis, especially in the context of sustainable water-related applications. Moreover, for polymeric photocatalysts, a comprehensive understanding of exciton dissociation and its impact on material characteristics is still lacking. The use of a donor–acceptor (D-A) system might potentially enhance exciton dissociation. The complete dissociation of excitons is facilitated by the additional energy produced during an excited donor electron's transfer to the acceptor. Interfacial dipoles and exciton delocalization may also encourage exciton dissociation.

A complicated set of photophysical and electrocatalytic processes are involved in the synthesis of solar fuel by photocatalysis. The six components of photocatalytic reactions are as follows: mass transfer of reactants and products, photon absorption, exciton separation, carrier diffusion, carrier transport, and catalytic efficiency. The fundamental efficiencies are impacted by these six characteristics in different ways. To obtain a high AQE, all of these characteristics need be optimized, even if they occur at various timescales and spatial resolutions. Addressing this topic seems to be necessary since different linear, planarized, non-crosslinked, and nonporous conjugated polymers have been reported to have photocatalytic activity that equals or exceeds that of microporous polymer photocatalysts. The processes involved in producing solar fuels through photocatalysis are both complex and sophisticated. The integration of photophysical and electrocatalytic reactions is crucial to developing efficient and sustainable technologies for solar-driven fuel generation. The impacts of the six steps of photocatalytic reactions on fundamental efficiency are diverse. Although they happen at different timescales and spatial resolutions, each of these features has to be adjusted to achieve high AQE values.

There are no standards by which to evaluate and contrast the different claimed photocatalyst efficiencies. Several parameters, including the sample mass, reactor setup, light source, and filters utilized, affect the overall amount of hydrogen created per unit of time. As such, when examined in different laboratories, the same material may yield different findings. The authors contend that stating the mass catalyst (mol h^{-1} g^{-1}) and the rate of hydrogen generation in units per time exacerbates the issue, since the mass of the catalyst and hydrogen production do not follow a linear relationship. Reporting AQE values presents a similar problem, as they are often not measured according to a strict protocol. The production of solar fuels provides opportunities to stop catastrophic climate change. It is essential to comprehend what functions and why in order to evaluate solar fuel initiatives. In order to arrive at a definitive answer, we must first apply our combined expertise to conduct a thorough global analysis of energy-innovation initiatives in order to identify optimal approaches.

References and further reading

[1] Lewis N S and Nocera D G 2006 Powering the planet: chemical challenges in solar energy utilization *Proc. Natl Acad. Sci. USA* **103** 15729–35

[2] Trancik J E 2014 Back the renewables boom *Nature* **507** 300–2

[3] Zou X and Zhang Y 2015 Noble metal-free hydrogen evolution catalysts for water splitting *Chem. Soc. Rev.* **44** 5148–80

[4] Cook T R, Dogutan D K, Reece S Y, Surendranath Y, Teets T S and Nocera D G 2010 Solar Energy supply and storage for the legacy and nonlegacy worlds *Chem. Rev.* **110** 6474–502

[5] Subbaraman R, Tripkovic D, Strmcnik D, Chang K C, Uchimura M, Paulikas A P, Stamenkovic V and Markovic N M 2011 Enhancing hydrogen evolution activity in water splitting by tailoring Li^+–$Ni(OH)_2$–Pt interfaces *Science (1979)* **334** 1256–60
[6] Xiao Y, Hu T, Zhao X, Hu F X, Yang H B and Li C M 2020 Thermo-selenizing to rationally tune surface composition and evolve structure of stainless steel to electrocatalytically boost oxygen evolution reaction *Nano Energy* **75** 104949
[7] Yao Q, DIng Y and Lu Z H 2020 Noble-metal-free nanocatalysts for hydrogen generation from boron- and nitrogen-based hydrides *Inorg. Chem. Front.* **7** 3837–74
[8] Chen S, Huang D, Xu P, Xue W, Lei L, Cheng M, Wang R, Liu X and Deng R 2020 Semiconductor-based photocatalysts for photocatalytic and photoelectrochemical water splitting: will we stop with photocorrosion? *J. Mater. Chem. A Mater.* **8** 2286–322
[9] Bonaccorso F, Colombo L, Yu G, Stoller M, Tozzini V, Ferrari A C, Ruoff R S and Pellegrini V 2015 Graphene, related two-dimensional crystals, and hybrid systems for energy conversion and storage *Science (1979)* **347** 1246501
[10] Jensen L (ed.) 2021 *The Sustainable Development Goals Report 2021* United Nations Statistics Division Development Data and Outreach Branch, New York https://unstats.un.org/sdgs/report/2021/
[11] Bonde J, Moses P G, Jaramillo T F, Nørskov J K and Chorkendorff I 2008 Hydrogen evolution on nano-particulate transition metal sulfides *Faraday Discuss.* **140** 219–31
[12] Mamiyev Z Q and Balayeva N O 2015 Preparation and optical studies of PbS nanoparticles *Opt. Mater. (Amst)* **46** 522–5
[13] Maeda K, Teramura K, Lu D, Takata T, Saito N, Inoue Y and Domen K 2006 Photocatalyst releasing hydrogen from water *Nature* **440** 295 295
[14] Balayeva N O and Mamiyev Z Q 2016 Synthesis and characterization of Ag_2S/PVA–fullerene (C_{60}) nanocomposites *Mater. Lett.* **175** 231–5
[15] Baran T, Wojtyła S, Dibenedetto A, Aresta M and Macyk W 2015 Zinc sulfide functionalized with ruthenium nanoparticles for photocatalytic reduction of CO_2 *Appl. Catal.* B **178** 170–6
[16] Ganapathy M, Chang C T and Alagan V 2022 Facile preparation of amorphous $SrTiO_3$-crystalline PbS heterojunction for efficient photocatalytic hydrogen production *Int. J. Hydrogen Energy* **47** 27555–65
[17] Reber J F and Meier K 1984 Photochemical production of hydrogen with zinc sulfide suspensions *J. Phys. Chem.* **88** 5903–13
[18] Keimer B and Moore J E 2017 The physics of quantum materials *Nat. Phys.* **13** 1045–55
[19] Sivula K and Van De Krol R 2016 Semiconducting materials for photoelectrochemical energy conversion *Nat. Rev. Mater.* **1** 15010
[20] Amirav L and Alivisatos A P 2010 Photocatalytic hydrogen production with tunable nanorod heterostructures *J. Phys. Chem. Lett.* **1** 1051–4
[21] Shiga Y, Umezawa N, Srinivasan N, Koyasu S, Sakai E and Miyauchi M 2016 A metal sulfide photocatalyst composed of ubiquitous elements for solar hydrogen production *Chem. Commun.* **52** 7470–3
[22] Hou H, Yuan Y, Cao S, Yang Y, Ye X and Yang W 2020 $CuInS_2$ nanoparticles embedded in mesoporous TiO_2 nanofibers for boosted photocatalytic hydrogen production *J. Mater. Chem. C Mater.* **8** 11001–7

[23] Caudillo-Flores U, Kubacka A, Berestok T, Zhang T, Llorca J, Arbiol J, Cabot A and Fernández-García M 2020 Hydrogen photogeneration using ternary $CuGaS_2$–TiO_2–Pt nanocomposites *Int. J. Hydrogen Energy* **45** 1510–20

[24] Singh J and Soni R K 2021 Enhanced sunlight driven photocatalytic activity of In_2S_3 nanosheets functionalized MoS_2 nanoflowers heterostructures *Sci. Rep.* **11** 14

[25] Armaroli N and Balzani V 2016 Solar electricity and solar fuels: status and perspectives in the context of the energy transition *Chem.—A Eur. J.* **22** 32–57

[26] Wang W, Yu J C, Xia D, Wong P K and Li Y 2013 Graphene and g-C 3 N 4 nanosheets cowrapped elemental α-sulfur as a novel metal-free heterojunction photocatalyst for bacterial inactivation under visible-light *Environ. Sci. Technol.* **47** 8724–32

[27] Markard J 2018 The next phase of the energy transition and its implications for research and policy *Nat. Energy* **3** 628–33

[28] Maeda K 2011 Photocatalytic water splitting using semiconductor particles: history and recent developments *J. Photochem. Photobiol.*, C **12** 237–68

[29] Rahman M Z, Kibria M G and Mullins C B 2020 Metal-free photocatalysts for hydrogen evolution *Chem. Soc. Rev.* **49** 1887–931

[30] Brousse T *et al* 2017 Materials for electrochemical capacitors *Springer Handbook of Electrochemical Energy* (Berlin: Springer) 495–561

[31] Turner J A 2004 Sustainable hydrogen production *Science (1979)* **305** 972–4

[32] Schlapbach L and Züttel A 2001 Hydrogen-storage materials for mobile applications *Nature* **414** 353–8

[33] Navarro R M, Peña M A and Fierro J L G 2007 Hydrogen production reactions from carbon feedstocks: fossil fuels and biomass *Chem. Rev.* **107** 3952–91

[34] Kumaravel V and Kang M 2020 Photocatalytic hydrogen evolution *Catalysts* **10** 6–7

[35] Fujishima A and Honda K 1972 Electrochemical photolysis of water at a semiconductor electrode *Nature* **238** 37–8

[36] Kumaravel V, Imam M D, Badreldin A, Chava R K, Do J Y, Kang M and Abdel-Wahab A 2019 Photocatalytic hydrogen production: role of sacrificial reagents on the activity of oxide, carbon, and sulfide catalysts *Catalysts* **9** 276

[37] Chen X, Shen S, Guo L and Mao S S 2010 Semiconductor-based photocatalytic hydrogen generation *Chem. Rev.* **110** 6503–70

[38] Kudo A and Miseki Y 2009 Heterogeneous photocatalyst materials for water splitting *Chem. Soc. Rev.* **38** 253–78

[39] Osterloh F E 2008 Inorganic materials as catalysts for photochemical splitting of water *Chem. Mater.* **20** 35–54

[40] Moriya Y, Takata T and Domen K 2013 Recent progress in the development of (oxy) nitride photocatalysts for water splitting under visible-light irradiation *Coord. Chem. Rev.* **257** 1957–69

[41] Hisatomi T, Kubota J and Domen K 2014 Recent advances in semiconductors for photocatalytic and photoelectrochemical water splitting *Chem. Soc. Rev.* **43** 7520–35

[42] Li X, Low J and Yu J 2016 Photocatalytic Hydrogen Generation *Photocatalysis: Applications* ed D D Dionysiou *et al* (London: The Royal Society of Chemistry) 10

[43] Chun W J, Ishikawa A, Fujisawa H, Takata T, Kondo J N, Hara M, Kawai M, Matsumoto Y and Domen K 2003 Conduction and valence band positions of Ta_2O_5, TaOn, and Ta_3N_5 by UPS and electrochemical methods *J. Phys. Chem.* B **107** 1798–803

[44] Hitoki G, Ishikawa A, Takata T, Kondo J N, Hara M and Domen K 2002 Ta_3N_5 as a novel visible light-driven photocatalyst ($\lambda < 600$nm) *Chem. Lett.* **31** 736–7
[45] Hitoki G, Takata T, Kondo J N, Hara M, Kobayashi H and Domen K 2002 An oxynitride, TaON, as an efficient water oxidation photocatalyst under visible light irradiation ($\lambda \leqslant 500$ nm) *Chem. Commun.* **2** 1698–9
[46] Maeda K, Takata T, Hara M, Saito N, Inoue Y, Kobayashi H and Domen K 2005 GaN: ZnO solid solution as a photocatalyst for visible-light-driven overall water splitting *J. Am. Chem. Soc.* **127** 8286–7
[47] Liu M, Jing D, Zhou Z and Guo L 2013 Twin-induced one-dimensional homojunctions yield high quantum efficiency for solar hydrogen generation *Nat. Commun.* **4** 2278
[48] Zhang K and Guo L 2013 Metal sulphide semiconductors for photocatalytic hydrogen production *Catal. Sci. Technol.* **3** 1672–90
[49] Tsuji I, Kato H, Kobayashi H and Kudo A 2004 Photocatalytic H_2 evolution reaction from aqueous solutions over band structure-controlled $(AgIn)_xZn_{2(1-x)}S_2$ solid solution photocatalysts with visible-light response and their surface nanostructures *J. Am. Chem. Soc.* **126** 13406–13
[50] Liao L *et al* 2014 Efficient solar water-splitting using a nanocrystalline CoO photocatalyst *Nat. Nanotechnol.* **9** 69–73
[51] Wang X, Maeda K, Thomas A, Takanabe K, Xin G, Carlsson J M, Domen K and Antonietti M 2009 A metal-free polymeric photocatalyst for hydrogen production from water under visible light *Nat. Mater.* **8** 76–80
[52] Cao S, Low J, Yu J and Jaroniec M 2015 Polymeric photocatalysts based on graphitic carbon nitride *Adv. Mater.* **27** 2150–76
[53] Wang X, Blechert S and Antonietti M 2012 Polymeric graphitic carbon nitride for heterogeneous photocatalysis *ACS Catal.* **2** 1596–606
[54] Rahman M Z and Edvinsson T 2019 What is limiting pyrite solar cell performance? *Joule* **3** 2290–3
[55] Rahman M Z and Edvinsson T 2019 How to make a most stable perovskite solar cell *Matter* **1** 562–4
[56] Pender J P, Guerrera J V, Wygant B R, Weeks J A, Ciufo R A, Burrow J N, Walk M F, Rahman M Z, Heller A and Mullins C B 2019 Carbon nitride transforms into a high lithium storage capacity nitrogen-rich carbon *ACS Nano* **13** 9279–91
[57] Linsebigler A L, Lu G and Yates J T 1995 Photocatalysis on TiO_2 surfaces: principles, mechanisms, and selected results *Chem. Rev.* **95** 735–58
[58] Chen X and Mao S S 2007 Titanium dioxide nanomaterials: synthesis, properties, modifications and applications *Chem. Rev.* **107** 2891–959
[59] Li R and Li C 2017 Photocatalytic Water Splitting on Semiconductor-Based Photocatalysts *Advances in Catalysis* ed C Song (Amsterdam: Elsevier) 60 1–57
[60] Serpone N 1997 Relative photonic efficiencies and quantum yields in heterogeneous photocatalysis *J. Photochem. Photobiol. A Chem.* **104** 1–12
[61] Sayama K and Arakawa H 1997 Effect of carbonate salt addition on the photocatalytic decomposition of liquid water over Pt–TiO_2 catalyst *J. Chem. Soc., Faraday Trans.* **93** 1647–54
[62] Ciamician G 1912 The photochemistry of the future *Science* **36** 385–94
[63] Kisch H 2013 Semiconductor photocatalysis—mechanistic and synthetic aspects *Angew. Chem. Int. Ed.* **52** 812–47

[64] Ravelli D, Dondi D, Fagnoni M and Albini A 2009 Photocatalysis. A multi-faceted concept for green chemistry *Chem. Soc. Rev.* **38** 1999–2011
[65] Fagnoni M, Dondi D, Ravelli D and Albini A 2007 Photocatalysis for the formation of the C–C bond *Chem. Rev.* **107** 2725–56
[66] Ohtani B 2010 Photocatalysis A to Z: what we know and what we do not know in a scientific sense *J. Photochem. Photobiol.,* C **11** 157–78
[67] Rahman M Z 2014 Advances in surface passivation and emitter optimization techniques of c-Si solar cells *Renew. Sustain. Energy Rev.* **30** 734–42
[68] Yue M, Lambert H, Pahon E, Roche R, Jemei S and Hissel D 2021 Hydrogen energy systems: a critical review of technologies, applications, trends and challenges *Renew. Sustain. Energy Rev.* **146** 111180
[69] Rahman M Z, Tang Y and Kwong P 2018 Reduced recombination and low-resistive transport of electrons for photo-redox reactions in metal-free hybrid photocatalyst *Appl. Phys. Lett.* **112** 253902
[70] Hisatomi T, Takanabe K and Domen K 2015 Photocatalytic water-splitting reaction from catalytic and kinetic perspectives *Catal. Lett.* **145** 95–108
[71] Rahman M Z and Mullins C B 2019 Understanding charge transport in carbon nitride for enhanced photocatalytic solar fuel production *Acc. Chem. Res.* **52** 248–57
[72] Kroeze J E, Savenije T J and Warman J M 2004 Electrodeless determination of the trap density, decay kinetics, and charge separation efficiency of dye-sensitized nanocrystalline TiO_2 *J. Am. Chem. Soc.* **126** 7608–18
[73] Enright B and Fitzmaurice D 1996 Spectroscopic determination of electron and hole effective masses in a nanocrystalline semiconductor film *J. Phys. Chem.* **100** 1027–35
[74] Maeda K 2013 Z-scheme water splitting using two different semiconductor photocatalysts *ACS Catal.* **3** 1486–503
[75] Chen S, Takata T and Domen K 2017 Particulate photocatalysts for overall water splitting *Nat. Rev. Mater.* **2** 17050
[76] Bard A J 1980 Photoelectrochemistry *Science (1979)* **207** 139–44
[77] Serpone N and Emeline A V 2012 Semiconductor photocatalysis—past, present, and future outlook *J. Phys. Chem. Lett.* **3** 673–7
[78] Fujishima A, Zhang X and Tryk D A 2008 TiO_2 photocatalysis and related surface phenomena *Surf. Sci. Rep.* **63** 515–82
[79] Sato J, Saito N, Yamada Y, Maeda K, Takata T, Kondo J N, Hara M, Kobayashi H, Domen K and Inoue Y 2005 RuO_2-loaded β-Ge_3N_4 as a non-oxide photocatalyst for overall water splitting *J. Am. Chem. Soc.* **127** 4150–1
[80] Arai N, Saito N, Nishiyama H, Inoue Y, Domen K and Sato K 2006 Overall water splitting by RuO_2-dispersed divalent-ion-doped GaN photocatalysts with d^{10} electronic configuration *Chem. Lett.* **35** 796
[81] Gueymard C A 2004 The sun's total and spectral irradiance for solar energy applications and solar radiation models *Sol. Energy* **76** 423–53
[82] Niishiro R, Kato H and Kudo A 2005 Nickel and either tantalum or niobium-codoped TiO_2 and $SrTiO_3$ photocatalysts with visible-light response for H_2 or O_2 evolution from aqueous solutions *Phys. Chem. Chem. Phys.* **7** 2241
[83] Sakata Y, Matsuda Y, Yanagida T, Hirata K, Imamura H and Teramura K 2008 Effect of metal ion addition in a Ni supported Ga_2O_3 photocatalyst on the photocatalytic overall splitting of H_2O *Catal. Lett.* **125** 22–6

[84] Asahi R, Morikawa T, Ohwaki T, Aoki K and Taga Y 2001 Visible-light photocatalysis in nitrogen-doped titanium oxides *Science (1979)* **293** 269–71
[85] Yashima M, Lee Y and Domen K 2007 Crystal structure and electron density of tantalum oxynitride, a visible light responsive photocatalyst *Chem. Mater.* **19** 588–93
[86] Li X, Kikugawa N and Ye J 2008 Nitrogen-doped lamellar niobic acid with visible light-responsive photocatalytic activity *Adv. Mater.* **20** 3816–9
[87] Fang J, Cao S W, Wang Z, Shahjamali M M, Loo S C J, Barber J and Xue C 2012 Mesoporous plasmonic Au–TiO_2 nanocomposites for efficient visible-light-driven photocatalytic water reduction *Int. J. Hydrogen Energy* **37** 17853–61
[88] Kowalska E, Mahaney O O P, Abe R and Ohtani B 2010 Visible-light-induced photocatalysis through surface plasmon excitation of gold on titania surfaces *Phys. Chem. Chem. Phys.* **12** 2344–55
[89] Méndez-Medrano M G, Kowalska E, Lehoux A, Herissan A, Ohtani B, Rau S, Colbeau-Justin C, Rodríguez-López J L and Remita H 2016 Surface modification of TiO_2 with Au nanoclusters for efficient water treatment and hydrogen generation under visible light *J. Phys. Chem.* C **120** 25010–22
[90] Bae E, Choi W, Park J, Shin H S, Kim S B and Lee J S 2004 Effects of surface anchoring groups (carboxylate vs phosphonate) in ruthenium-complex-sensitized TiO_2 on visible light reactivity in aqueous suspensions *J. Phys. Chem.* B **108** 14093–101
[91] Darwent J R and Mills A 1982 Photo-oxidation of water sensitized by WO_3 powder *J. Chem. Soc., Faraday Trans. 2* **78** 359–67
[92] Erbs W, Desilvestro J, Borgarello E and Grätzel M 1984 Visible-light-induced O_2 generation from aqueous dispersions of WO_3 *J. Phys. Chem.* **88** 4001–6
[93] Scaife D E 1980 Oxide semiconductors in photoelectrochemical conversion of solar energy *Sol. Energy* **25** 41–54
[94] Kato H and Kudo A 2002 Visible-light-response and photocatalytic activities of TiO_2 and $SrTiO_3$ photocatalysts codoped with antimony and chromium *J. Phys. Chem.* B **106** 5029–34
[95] Kim H G, Hwang D W and Lee J S 2004 An undoped, single-phase oxide photocatalyst working under visible light *J. Am. Chem. Soc.* **126** 8912–3
[96] Hosogi Y, Shimodaira Y, Kato H, Kobayashi H and Kudo A 2008 Role of Sn^{2+} in the band structure of SnM_2O_6 and $Sn_2M_2O_7$ (M = Nb and Ta) and their photocatalytic properties *Chem. Mater.* **20** 1299–1307
[97] Abe R, Sayama K and Arakawa H 2003 Significant influence of solvent on hydrogen production from aqueous I_3^-/I^- redox solution using dye-sensitized Pt/TiO_2 photocatalyst under visible light irradiation *Chem. Phys. Lett.* **379** 230–5
[98] Cargnello M, Gasparotto A, Gombac V, Montini T, Barreca D and Fornasiero P 2011 Photocatalytic H_2 and added-value by-products-the role of metal oxide systems in their synthesis from oxygenates *Eur. J. Inorg. Chem.* **2011** 4309–23
[99] Simon Q *et al* 2011 Plasma-assisted synthesis of Ag/ZnO nanocomposites: first example of photo-induced H_2 production and sensing *Int. J. Hydrogen Energy* **36** 15527–37
[100] Gasparotto A *et al* 2011 F-doped Co_3O_4 photocatalysts for sustainable H_2 generation from water/ethanol *J. Am. Chem. Soc.* **133** 19362–5
[101] Simon Q, Barreca D, Gasparotto A, MacCato C, Montini T, Gombac V, Fornasiero P, Lebedev O I, Turner S and Van Tendeloo G 2012 Vertically oriented CuO/ZnO nanorod

arrays: from plasma-assisted synthesis to photocatalytic H_2 production *J. Mater. Chem.* **22** 11739–47

[102] Zheng N, Bu X, Vu H and Feng P 2005 Open-framework chalcogenides as visible-light photocatalysts for hydrogen generation from water *Angew. Chem. Int. Ed.* **44** 5299–303

[103] Zheng N, Bu X and Feng P 2005 $Na_5(In_4S)(InS_4)_3{\cdot}6H_2O$, a zeolite-like structure with unusual SIn_4 tetrahedra *J. Am. Chem. Soc.* **127** 5286–7

[104] Tsuji I, Kato H and Kudo A 2005 Visible-light-induced H_2 evolution from an aqueous solution containing sulfide and sulfite over a ZnS–$CuInS_2$–$AgInS_2$ solid-solution photocatalyst *Angew. Chem. Int. Ed.* **44** 3565–8

[105] Acar C, Dincer I and Zamfirescu C 2014 A review on selected heterogeneous photocatalysts for hydrogen production *Int. J. Energy Res.* **38** 1903–20

[106] Kakuta N, Park K H, Finlayson M F, Ueno A, Bard A J, Campion A, Fox M A, Webber S E and White J M 1985 Photoassisted hydrogen production using visible light and coprecipitated ZnS·CdS without a noble metal *J. Phys. Chem.* **89** 732–4

[107] Shangguan W and Yoshida A 2002 Photocatalytic hydrogen evolution from water on nanocomposites incorporating cadmium sulfide into the interlayer *J. Phys. Chem.* B **106** 12227–30

[108] Zhang J, Chen X, Takanabe K, Maeda K, Domen K, Epping J D, Fu X, Antonieta M and Wang X 2010 Synthesis of a carbon nitride structure for visible-light catalysis by copolymerization *Angew. Chem.—Int. Ed.* **49** 441–4

[109] Hara M, Takata T, Kondo J N and Domen K 2004 Photocatalytic reduction of water by TaON under visible light irradiation *Catal. Today* **90** 313–7

[110] Kessler F K, Zheng Y, Schwarz D, Merschjann C, Schnick W, Wang X and Bojdys M J 2017 Functional carbon nitride materials-design strategies for electrochemical devices *Nat. Rev. Mater.* **2** 17030

[111] Liu G, Niu P and Cheng H M 2013 Visible-light-active elemental photocatalysts *ChemPhysChem.* **14** 885–92

[112] Peng C, Gao J, Wang S, Zhang X, Zhang X and Sun X 2011 Stability of hydrogen-terminated surfaces of silicon nanowires in aqueous solutions *J. Phys. Chem.* C **115** 3866–71

[113] Wang F, Ng W K H, Yu J C, Zhu H, Li C, Zhang L, Liu Z and Li Q 2012 Red phosphorus: an elemental photocatalyst for hydrogen formation from water *Appl. Catal.* B **111–2** 409–14

[114] Liu G, Niu P, Yin L and Cheng H M 2012 α-sulfur crystals as a visible-light-active photocatalyst *J. Am. Chem. Soc.* **134** 9070–3

[115] Chiou Y D and Hsu Y J 2011 Room-temperature synthesis of single-crystalline Se nanorods with remarkable photocatalytic properties *Appl. Catal.* B **105** 211–9

[116] Wang Y *et al* 2019 Current understanding and challenges of solar-driven hydrogen generation using polymeric photocatalysts *Nat. Energy* **4** 746–60

[117] Ghosh S, Kouamé N A, Ramos L, Remita S, Dazzi A, Deniset-Besseau A, Beaunier P, Goubard F, Aubert P H and Remita H 2015 Conducting polymer nanostructures for photocatalysis under visible light *Nat. Mater.* **14** 505–11

[118] Ghosh S, Kouame N A, Remita S, Ramos L, Goubard F, Aubert P H, Dazzi A, Deniset-Besseau A and Remita H 2015 Visible-light active conducting polymer nanostructures with superior photocatalytic activity *Sci. Rep.* **5** 18002

[119] Yuan X, Dragoe D, Beaunier P, Uribe D B, Ramos L, Méndez-Medrano M G and Remita H 2020 Polypyrrole nanostructures modified with mono- and bimetallic nanoparticles for photocatalytic H_2 generation *J. Mater. Chem.* A **8** 268–77

[120] Kato H, Asakura K and Kudo A 2003 Highly efficient water splitting into H_2 and O_2 over lanthanum-doped $NaTaO_3$ photocatalysts with high crystallinity and surface nanostructure *J. Am. Chem. Soc.* **125** 3082–9

[121] Sakata Y, Hayashi T, Yasunaga R, Yanaga N and Imamura H 2015 Remarkably high apparent quantum yield of the overall photocatalytic H_2O splitting achieved by utilizing Zn ion added Ga_2O_3 prepared using dilute $CaCl_2$ solution *Chem. Commun.* **51** 12935–8

[122] Ham Y, Hisatomi T, Goto Y, Moriya Y, Sakata Y, Yamakata A, Kubota J and Domen K 2016 Flux-mediated doping of $SrTiO_3$ photocatalysts for efficient overall water splitting *J. Mater. Chem.* A **4** 3027–33

[123] Zou Z, Ye J, Sayama K and Arakawa H 2001 Direct splitting of water under visible light irradiation with an oxide semiconductor photocatalyst *Nature* **414** 625–7

[124] Pan C, Takata T, Nakabayashi M, Matsumoto T, Shibata N, Ikuhara Y and Domen K 2015 A complex perovskite-type oxynitride: the first photocatalyst for water splitting operable at up to 600 nm *Angew. Chem.—Int. Ed.* **54** 2955–9

[125] Yeh T F, Teng C Y, Chen S J and Teng H 2014 Nitrogen-doped graphene oxide quantum dots as photocatalysts for overall water-splitting under visible light illumination *Adv. Mater.* **26** 3297–303

[126] Kibria M G, Nguyen H P T, Cui K, Zhao S, Liu D, Guo H, Trudeau M L, Paradis S, Hakima A R and Mi Z 2013 One-step overall water splitting under visible light using multiband InGaN/GaN nanowire heterostructures *ACS Nano* **7** 7886–93

[127] Zhang C, Chen C, Dong H, Shen J R, Dau H and Zhao J 2015 A synthetic Mn_4Ca-cluster mimicking the oxygen-evolving center of photosynthesis *Science (1979)* **348** 690–3

[128] Wang W, Chen J, Li C and Tian W 2014 Achieving solar overall water splitting with hybrid photosystems of photosystem II and artificial photocatalysts *Nat. Commun.* **5** 4647

[129] Liu G, Ye S, Yan P, Xiong F, Fu P, Wang Z, Chen Z, Shi J and Li C 2016 Enabling an integrated tantalum nitride photoanode to approach the theoretical photocurrent limit for solar water splitting *Energy Environ. Sci.* **9** 1327–34

Chapter 6

Photocatalysis for organic degradation using perovskite materials

Phuong Hoang Nguyen, Thi Minh Cao and Viet Van Pham

6.1 Introduction

In recent years, the issue of organic compound pollution, especially that caused by persistent organic pollutants (POPs), has become increasingly severe [1, 2]. These compounds are often used in producing various industrial products, leading to their widespread release into the environment. Organic compounds can be found in multiple water sources, including rivers, lakes, and groundwater. They can also enter the food chain through the consumption of contaminated water, causing harm to human and animal health [3]. In addition, long-term exposure to organic compounds can lead to a variety of health problems, including cancer, reproductive disorders, and immune system damage [4]. POPs and industrial chemicals that were manufactured in great quantities during the 20th century directly affect human and animal health. When examining the effects of pollution on morbidity and mortality in relation to other risk factors, the results indicate that pollution remains one of the most significant contributors to disease and premature death worldwide. According to figure 6.1, the number of deaths attributed to pollution is comparable to the number of deaths caused by smoking. This highlights the substantial and ongoing impact of pollution on global health outcomes. Exposure to high levels of persistent organic pollutants (POPs) can have detrimental health effects on both humans and animals. These effects are well documented and may include various adverse outcomes, such as cancer, nervous system damage, reproductive disorders, and immune system disruption. It is important to recognize that POPs have the potential to cause significant harm to living organisms, highlighting the need for effective measures to reduce exposure and mitigate their impact on health [5].

To address this issue, it is necessary to implement effective wastewater treatment processes in production facilities, enhance the monitoring and assessment of water quality to detect pollutants promptly, and implement other measures such as the use

doi:10.1088/978-0-7503-5697-8ch6

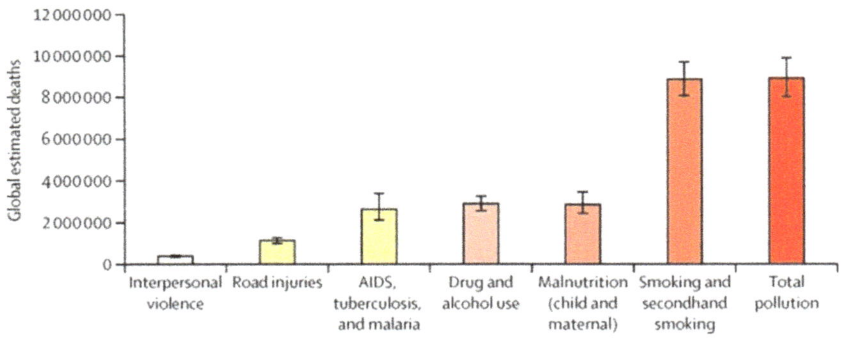

Figure 6.1. Global estimated deaths by major risk factor or cause. Reprinted from The Lancet [6], Copyright (2022), with permission from Elsevier.

of wastewater treatment technologies, a reduction in the chemicals used in production, and the use of recycled water sources. Furthermore, increasing education and awareness about environmental pollution is also critical in reducing pollution. Governments and nongovernmental organizations need to improve the information, education, and training offered to communities to raise awareness about environmental issues and enhance cooperation and consultation with businesses and organizations in implementing effective pollution reduction measures.

In terms of technical approaches, several methods can be used to reduce organic pollutants, which can be broadly categorized into prevention, treatment, and remediation approaches. For instance, biological treatment based on the use of microorganisms to degrade organic pollutants; physical and chemical treatment; phytoremediation; and adsorption. Therein, advanced oxidation processes (AOPs) can be employed to remove or degrade organic pollutants. These include activated carbon adsorption, advanced oxidation processes (e.g. ozonation, UV irradiation, photocatalysis, and chemical oxidation), air stripping, and membrane filtration. Among advance oxidation methods, photocatalysis and other processes based on photocatalysis, i.e. the photo-Fenton method, the use of a peroxymonosulfate (PMS) activator, etc. are promising technologies for the treatment of water contaminated with organic compounds. These processes use light to activate a photocatalyst, typically TiO_2, to generate highly reactive oxygen species that can degrade organic pollutants. One of the significant advantages of photocatalysis is that it is a green technology that does not require chemicals or produce harmful byproducts. In addition, photocatalysis can treat a wide range of organic contaminants, including pesticides, pharmaceuticals, and dyes. Furthermore, photocatalysis is a relatively simple and low-cost technology that can easily be scaled up for industrial applications. Overall, the use of photocatalysis in treating water contaminated with organic compounds offers several advantages over traditional treatment methods and shows great potential for widespread adoption in the future [7].

Perovskite materials have recently attracted significant attention in photocatalysis due to their unique optical and electronic properties [8]. These materials have a perovskite crystal structure that consists of a metal cation, an organic or inorganic cation, and an anion. Perovskite materials have a high absorption coefficient, excellent charge carrier mobility, and long carrier lifetimes, which make them ideal for photocatalytic applications. Several studies have investigated the application of perovskite materials in photocatalysis to degrade organic pollutants, split water, and reduce carbon dioxide. For example, researchers have reported the successful use of perovskite materials in the photocatalytic degradation of organic dyes and the production of hydrogen gas from water [7]. These promising results suggest that perovskite materials have the potential to become a new class of photocatalysts for a wide range of environmental and energy applications.

This chapter provides a complete overview of the concept, properties, and application of perovskites as photocatalysts for organic pollutant removal. The primary methods used to synthesize perovskites are summarized in this chapter. In addition, this chapter describes some of the most important applications of perovskites in the photocatalysis field, including photocatalysis, the photo-Fenton reaction, and PMS activation for the degradation of organic pollutants in wastewater. Some conclusions and prospects are also briefly introduced, which could lead to breakthrough developments in perovskite materials design.

6.2 An overview of perovskite materials

6.2.1 An introduction to perovskites

Perovskites, named after Russian mineralogist Lev Perovski, are compounds with a crystal structure that follows the formula ABX_3, where A and B are cations and X is an anion (figure 6.2). The perovskite structure is a crystal structure that was first discovered in the mineral calcium titanate ($CaTiO_3$). It is characterized by a three-dimensional network of corner-sharing BX_6 octahedra that form a cubic lattice. The A cation occupies the center of the cube and is surrounded by octahedra.

The arrangement of the ideal perovskite structure can be described as follows:

$$(r_A + r_O) = \frac{\sqrt{2}}{2}A = \sqrt{2}(r_A + r_O), \tag{6.1}$$

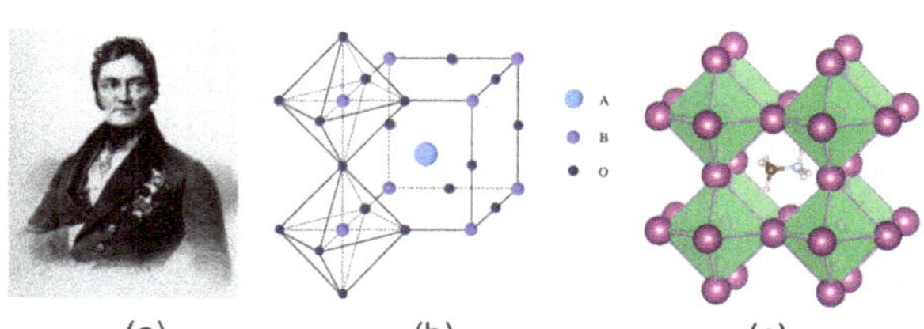

(a) (b) (c)

Figure 6.2. (a) Russian mineralogist Lev Perovski (1792–1856). This Count Lev Alekseevich Perovski image has been obtained by the author(s) from the Wikimedia website, where it is stated to have been released into the public domain. It is included within this book on that basis. (b) The structure of ABO_3 perovskite-type oxides. Reprinted from [10], Copyright (2021), with permission from Elsevier. (c) A methylammonium cation ($CH_3NH_3^+$) occupies the central A site, surrounded by 12 nearest-neighbor iodide ions in corner-sharing PbI_6 octahedra. Reproduced from [11]. CC BY 4.0.

where r_A, r_B, and r_O represent the ionic radii of the cations A and B and the anion O, respectively. Although the idealized structure is primitive cubic, differences in the sizes of the cations can cause distortions. This typically involves tilting of the octahedral units (known as octahedral tilting). The stability of the perovskite structure is described by a tolerance factor t, which represents the range of relative ionic sizes [3]:

$$t = \frac{(r_A + r_O)}{\sqrt{2}(r_B + r_O)}. \tag{6.2}$$

Empirically, if the tolerance factor is between 0.95 and 1.0, the structure is cubic. If $0.8 < t < 0.9$, the compound transforms into a tetragonal, rhombohedral, or orthorhombic crystal [9]. In photocatalytic studies, lattice distortion significantly affects the crystal field, altering dipoles and electronic band structures, which in turn impact the behaviors of photogenerated charge carriers, including excitation, transfer, and redox reactions, during the photocatalytic process.

Figure 6.3 shows that energy-dispersive x-ray (EDX) mapping of the elements in perovskite oxides ($BaTiO_3$ and $SrTiO_3$) exhibits visible particles on the particle clusters, and it also shows a uniform arrangement of Mo and S atoms on those clusters. $SrTiO_3$ and $BaTiO_3$ perovskite-type oxides offer a viable alternative to the predominantly used TiO_2 due to their favorable band positions, optical

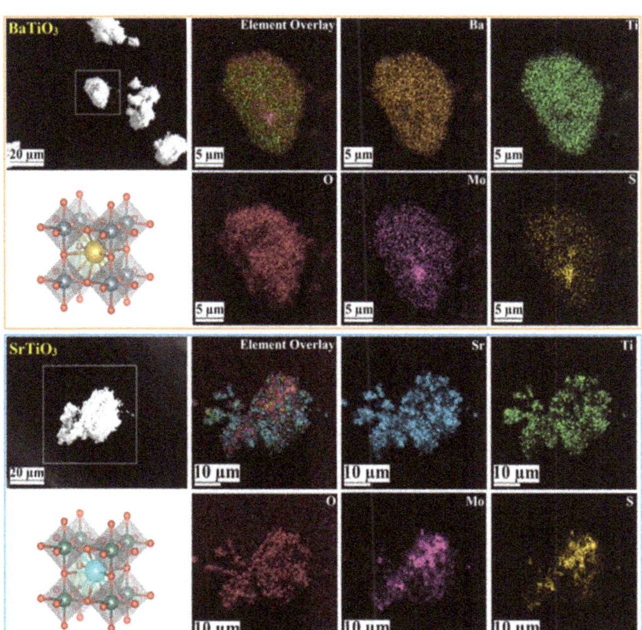

Figure 6.3. Elemental mapping images of photodeposited $BaTiO_3$ and $SrTiO_3$ nanocrystal clusters. Reprinted from [12], Copyright (2023), with permission from Elsevier.

characteristics, and crystallographic properties. These qualities make them suitable for effective photochemical energy conversion reactions.

In addition to the oxide-based perovskites, another type called 'perovskite halides' is a group of materials with the formula ABX_3, where A represents an organic and/or inorganic cation with a single oxidation state, B represents a group 4A metal with a valence of two (e.g. Pb^{2+}, Ge^{2+}, Sn^{2+}, etc.), and X corresponds to a halide anion, e.g. I^-, Cl^-, Br^-, etc [8]. These materials have gained tremendous attention due to their excellent optoelectronic properties, such as high light absorption coefficients, long carrier diffusion lengths, and high charge carrier mobilities.

In general, perovskites have shown great promise as low-cost and high-performance materials for solar cells. They have achieved remarkable power conversion efficiencies (PCEs) that exceed 25% and are comparable to the efficiencies of traditional silicon solar cells. In addition to their use in photovoltaics, metal halide perovskites have also been investigated for use in other optronics applications (light-emitting diodes (LEDs), photodetectors, photodiodes, and lasers). Overall, metal halide perovskites represent a promising class of materials with significant potential for advancing the field of solar energy and optoelectronics. Ongoing research and development efforts aim to address their stability issues and further enhance their performance for practical applications.

The bandgap energy of halide perovskites is one of the important factors that affect their optical properties. The bandgap is the energy required to push an electron from the valence band (VB) to the conduction band (CB), creating an electron–hole pair. The smaller the bandgap, the greater the semiconductor's ability to absorb light. Figure 6.4 shows the bandgaps of oxide-type perovskites (a) and halide-type perovskites (b) with the redox levels for the formation of reactive oxygen species (ROS) and $OH^\bullet$ radicals. For organic–inorganic perovskites ($MAPbI_3$), research studies have observed a small bandgap (about 1.5 eV) and a high light absorption coefficient of 105 cm^{-1} [13]. This means that $MAPbI_3$ one μm thick can absorb all light with a wavelength shorter than 800 nm in the solar spectrum. Moreover, by modifying factors such as size, composition substitution, phase segregation, phase transition, and deformation, it is possible to extensively adjust the range of light absorption in perovskite halides from UV to near-infrared wavelengths. The halogen anions that appear in the chemical formula $APbX_3$ for halide perovskites affect their CB energy levels. For instance, the bandgap of $MAPbI_3$ can be extensively tuned simply by changing the ratio of the different halogen ions. Similarly, continuous tuning of the spectrum can be achieved in inorganic halogen perovskites, as shown, for example, by $CsPbBr_3$, whose bandgap can be tuned from 1.88 to 3.03 eV by partially replacing Br with Cl or I. Furthermore, perovskite halides are a group of materials with unique photocatalytic properties. Their CBM and VBM positions are suitable for promoting many photocatalytic reactions [13].

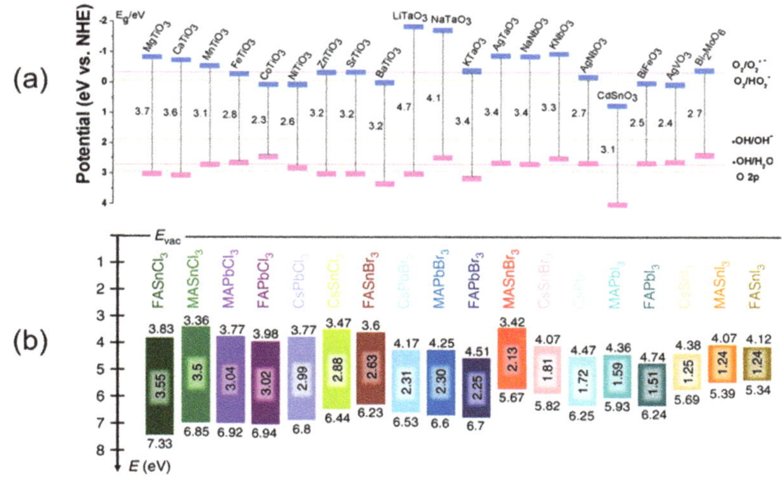

Figure 6.4. (a) E_g values (in eV) and the positions of the CB (upper) and the VB (lower) for various perovskite semiconductors at a pH of 7 vs. normal hydrogen electrode (NHE). Reprinted from [14], Copyright (2019), with permission from Elsevier. (b) A schematic energy level diagram of the 18 metal halide perovskites. Reproduced from [15]. CC BY 4.0.

6.2.2 The properties of perovskites

Perovskites have many advantageous properties for photovoltaics and photocatalysis due to their outstanding properties (figure 6.5).

6.2.2.1 Optoelectronic properties

Perovskite materials exhibit exceptional optoelectronic properties, making them suitable for various applications. Deschler *et al* [16] investigated the optoelectronic properties of perovskite materials and demonstrated that perovskite materials possess a high absorption coefficient, enabling efficient light harvesting across a wide spectrum, including visible and near-infrared wavelengths. In addition, they observed efficient charge transport due to long carrier diffusion lengths in these materials, making them suitable for application in solar cells and photocatalysis [17].

6.2.2.2 Tunable bandgap

Perovskite materials have quite diverse bandgap values that range from 2.1 to 4.0 eV (table 6.1) and a specific crystal structure that corresponds to a three-dimensional arrangement of metal cations, such as lead (Pb), tin (Sn), or other elements, surrounded by an octahedral arrangement of oxygen anions. Perovskite materials possessing the ABO_3 structure offer numerous benefits that make them a highly promising option for photocatalysis. These advantages include straightforward synthesis methods, adjustable physicochemical properties, well-suited frameworks, diverse levels of lattice distortion, and localized sites for surface oxidation and reduction reactions. These characteristics contribute to their potential to be excellent choices for photocatalytic processes.

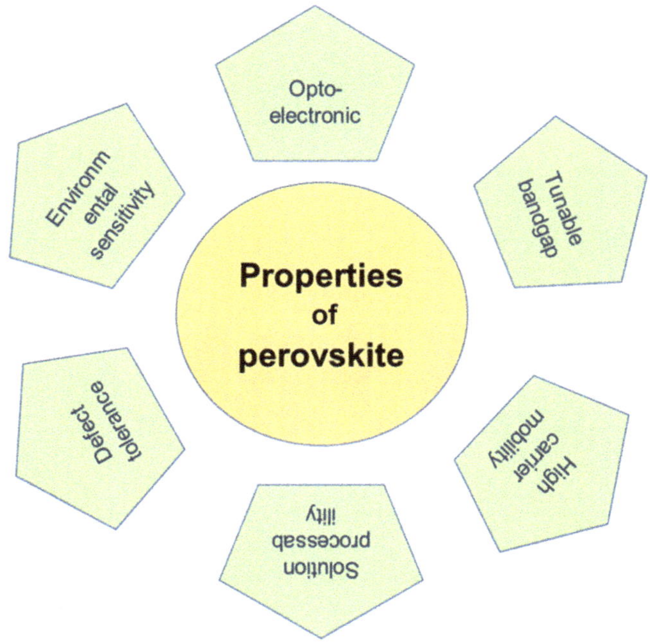

Figure 6.5. Some properties of perovskites.

Table 6.1. The bandgaps of some perovskite materials. Reproduced from [19]. CC BY 4.0.

Material	Bandgap (eV)
$SrTiO_3$	3.20
$NaTaO_3$	4.00
$CaTiO_3$	3.62
$BiFeO_3$	2.40
$LaFeO_3$	2.00
$NaNbO_3$	3.48
$LaCoO_3$	2.10
Bi_2WO_6	2.70
$La_2Ti_2O_7$	3.28

One of the remarkable features of perovskite materials is their tunable bandgap, which can be adjusted to absorb light across a wide range of energies. This tunability is advantageous for applications such as solar cells and light-emitting devices. Prasanna *et al* synthesized and investigated [18] tin and lead iodide perovskite semiconductors with the composition AMX_3, where M is a metal and X is a halide. The introduction of a smaller A-site cation into a perovskite structure can induce two distinct distortions: tilting of the MX_6 octahedra or isotropic

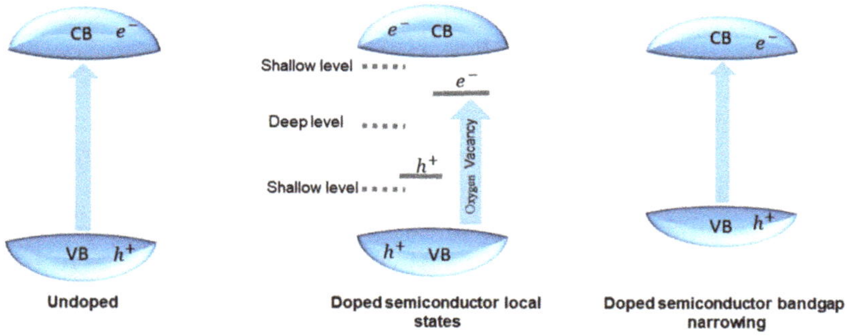

Figure 6.6. The effect of doping on the bandgap. Reproduced from [19]. CC BY 3.0.

contraction of the lattice. The former effect generally increases the bandgap, while the latter decreases it. In the case of lead iodide perovskites, replacing the larger formamidinium cation with the smaller cesium cation resulted in an expanded bandgap, which was attributed to octahedral tilting. Conversely, tin-based perovskites, being slightly smaller than lead, displayed no octahedral tilting upon cesium substitution but rather lattice contraction, leading to a progressive reduction in the bandgap. This study demonstrated a systematic approach to fine-tuning the bandgap and band positions of metal halide perovskites by controlling the cation composition. This research also showcased the development of solar cells that have bandgaps optimized for harvesting infrared light at up to 1040 nm, achieving a stabilized power conversion efficiency of 17.8%. The findings provide insights into cation-based bandgap tuning in 3D metal halide perovskites and have implications for the development of perovskite semiconductors for optoelectronic applications [18].

In addition, the control of the band positions and bandgaps of perovskites is also of interest in the visible-light-driven photocatalysis field. Figure 6.6 shows the effects on the bandgap of perovskites of doping to form new states, i.e. shallow and deep levels or oxygen vacancy states. Besides, the narrow bandgap of perovskites helps them become more effective in photocatalytic reactions under visible light conditions.

6.2.2.3 High carrier mobility

Perovskite materials exhibit high charge carrier mobility, allowing efficient movement of electrons and holes through the material. This high mobility is crucial to obtain high-performance electronic and optronics devices. Liu *et al* [20] synthesized InP and InAs nanocrystals (NCs) and used various inorganic ligands (molecular metal chalcogenide complexes (MCCs) and chalcogenide ions to cap the NCs). These inorganic ligands effectively replaced the organic ligands on the NC surfaces and stabilized the colloidal solutions of InP and InAs NCs in polar solvents. Moreover, they facilitated efficient charge transport between individual NCs. The films of InP and InAs NCs capped with MCCs exhibited high electron mobility, with $Cu_7S_4^-$ MCC ligands achieving electron mobility exceeding 15 $cm^2\ V^{-1}\ s^{-1}$. In addition, Lim *et al* [21] observed similar long-range mobilities of 0.3 to 6.7 $cm^2\ V^{-1}\ s^{-1}$ in perovskites such as $FA_{0.83}Cs_{0.17}Pb(I_{0.9}Br_{0.1})_3$, $(FA_{0.83}MA_{0.17})_{0.95}Cs_{0.05}Pb(I_{0.9}Br_{0.1})_3$, and

$CH_3NH_3PbI_{3-x}Cl_x$ polycrystalline films, highlighting their potential for use in various optoelectronic applications in which efficient charge transport is crucial.

6.2.2.4 Processability in solution

Perovskite materials can be processed in solution, which offers advantages for large-scale manufacturing processes. They can be deposited using techniques such as spin-coating, inkjet printing, or spray coating, enabling the fabrication of thin films with good uniformity and low-cost manufacturing potential. A study by Hu focused on scaling up perovskite photovoltaic technology through the exploration of film formation, device architectures, deposition methods, and scalable manufacturing approaches. This study emphasized the importance of precise control over nucleation and crystal growth during the film formation process to achieve high-performance perovskite solar cells [22].

6.2.2.5 Defect tolerance

Perovskite materials show a certain degree of tolerance toward defects, which can be beneficial for photocatalytic performance. Defects in the crystal structure can trap charge carriers, reducing their mobility. However, in perovskites, these trapped carriers can be released and continue to contribute to the device's functionality. Steirer *et al* [23] explored the defect tolerance characteristic of perovskite materials, with a specific focus on methylammonium lead triiodide ($MAPbI_3$) perovskite, and indicated that defects in the crystal structure influenced the performance and stability of energy-related devices utilizing this material. Furthermore, MAPbI3 perovskite exhibited a certain level of tolerance towards specific defects in its crystal structure. These defects have the potential to trap charge carriers, resulting in a decrease in their mobility. However, perovskite materials possess the capability to release these trapped charge carriers, enabling them to continue contributing to the device's overall performance [24].

Defect engineering provides a convenient and effective approach for improving the photocatalytic efficiency of perovskite materials. The substitution of different materials can introduce defects into the perovskite structure. These defects serve as sites that can trap charge carriers, leading to enhanced charge separation and increased photocatalytic efficiency. In addition, the existence of the intrinsic factors, e.g. Sn^{4+} that are easy formation of Sn vacancies, impacting to the defect chemistry and subsequent recombination of charge carriers, prolonged exposure of these materials to extrinsic factors (e.g., air, O_2) can severely degrade their key optoelectronic properties [17].

6.2.2.6 Environmental sensitivity

Perovskite materials can be sensitive to moisture, heat, and light exposure. They tend to degrade in the presence of moisture and are susceptible to thermal stress. However, ongoing research aims to improve their stability and durability for practical applications. Ahangharnejhad *et al* presented a method for protecting metal halide perovskite solar cells from degradation in high-humidity environments using a sputtered SiO_2 barrier coating. The SiO_2 protective layer significantly enhanced the devices' resistance to extreme humidity conditions, increasing their lifetimes by approximately 60 times for $CH_3NH_3PbI_3$ and around 600 times for triple-cation perovskite material. Real-time laser-beam-induced current (LBIC) measurements validated the effectiveness of this approach in preventing degradation at the edges of scribed lines, making it suitable for the production of monolithically integrated modules [25].

6.3 Methods used to synthesize perovskite materials

There are several methods for making perovskite materials, i.e. wet chemical methods (sol–gel process hydrothermal route, coprecipitation method, electrochemical synthesis, and emulsion or microemulsion synthesis, etc.), solid-state reaction methods, template-assisted methods, etc. These methods can be adjusted to control the properties of the perovskite material, such as its crystal structure, morphology, and surface area, which can affect its photocatalytic activity [26].

6.3.1 Wet chemical methods

Wet chemical synthesis, also known as solution-based synthesis, is a broad term that refers to a wide range of chemical processes used to create or modify materials in a liquid or dissolved form. It involves the reaction of chemical precursors dissolved in a solvent to form a desired product. The reaction can be carried out at various temperatures, pressures, and reaction times, depending on the specific synthesis requirements. The reactants may undergo various chemical transformations, such as precipitation, hydrolysis, redox reactions, or complexation, to form the desired product. Wet chemical synthesis is widely used in various fields, including chemistry, materials science, nanotechnology, and pharmaceuticals. It offers several advantages, such as simplicity, scalability, and the ability to control the composition, size, shape, and structure of the resulting materials.

6.3.1.1 The sol–gel method

The sol–gel technique is commonly used to prepare nanosized perovskites with a high surface area. It involves the mixture of precursor compounds in specific solvents to form a stable 'sol,' which is then transformed into a gel-like system (both liquid and solid phases). The gel is then dried to evaporate the solvent and undergoes heating to obtain the desired perovskite materials [27]. Parida *et al* [28] synthesized $LaFeO_3$ perovskite nanomaterials using the sol–gel method at 130 °C. This heating process caused changes in the color and viscosity of the sol. When further heated, the brown, porous dry gel automatically ignited due to the thermal redox reaction. Finally, a solid was created by a self-burning process and then activated at different temperatures (500 to 900 °C) for 2 h to obtain a visible-light photocatalytic material. Wang *et al* [29] prepared perovskite materials including $La(NO_3)_3$ and $B(NO_3)_2$ (B = Mn, Fe, Co, Ni, Cu, Zn) by a sol–gel method using an ethylene glycol and alcohol mixture as a complex. In general, this synthesis method is quite simple: drying at 100 °C and annealing and sintering processes (500 °C–700 °C) were performed to obtain the final product.

6.3.1.2 The hydrothermal method

The hydrothermal method is one of the standard techniques used to synthesize perovskite particles with controlled shapes. This method subjects a reaction mixture to high-temperature and -pressure conditions in a stainless steel autoclave, and the final product is collected after cooling. Zhao *et al* [30] synthesized $CaTiO_3$ particles with a controlled morphology. Their study pointed out that rectangular $CaTiO_3$ particles exhibited the highest photocatalytic activity for the degradation of methylene blue under UV irradiation due to their high specific surface area compared to the surface areas of cubic and spherical particles.

6.3.1.3 The coprecipitation method

The coprecipitation method is another important technique used to synthesize perovskite materials. It is conducted using a suitable precipitating agent to coprecipitate two or more precursor components homogeneously [31]. Moshtaghi *et al* synthesized $CaSnO_3$ nanocubes using a simple and green coprecipitation method [32]. To obtain the final product, $CaSnO_3$, the researchers utilized a complex precursor called $Ca(sal)_2$ as a calcium source and $SnCl_2$ as a source of tin. The synthesis procedure employed water as an environmentally friendly solvent, and various alkaline agents were used as precipitating agents. After the reaction, the resulting precipitate was subjected to centrifugation, followed by washing and drying. Subsequently, the obtained material was annealed at a temperature of 900 °C for 5 h. Using this process, the researchers successfully obtained the desired final product, $CaSnO_3$. Junwu *et al* [33] synthesized $LaCoO_3$ nanocrystals using the coprecipitation method, in which NaOH was used as the precipitating agent. After precipitation, the resulting precipitates were centrifuged, washed, and dried at 100 °C. The dried residue was then annealed at 600 °C to obtain the product.

Figure 6.7 illustrates the general method of precipitation synthesis. The precursors A and B are dissolved in solvents and then precipitated using a NaOH solution

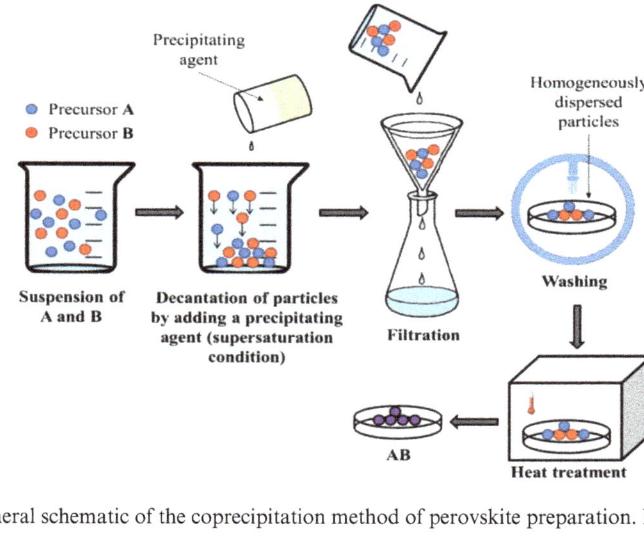

Figure 6.7. A general schematic of the coprecipitation method of perovskite preparation. Reprinted from [31], Copyright (2016), with permission from Elsevier.

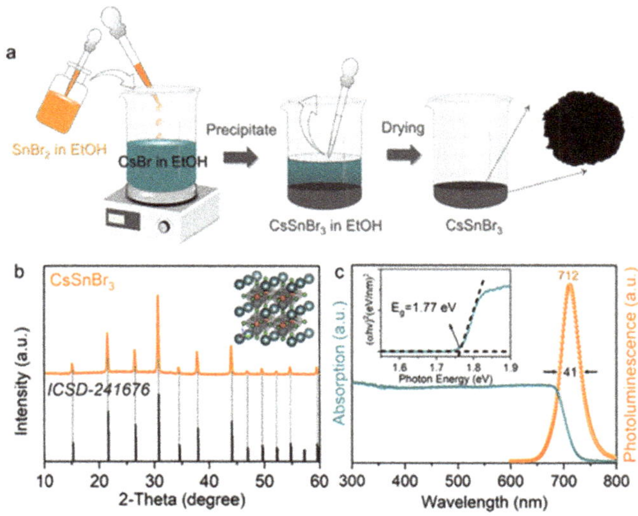

Figure 6.8. (a) A schematic of the coprecipitation process used to prepare $CsSnBr_3$ perovskite. (b) The x-ray diffraction pattern of $CsSnBr_3$. (c) The UV–vis absorption and PL spectra of $CsSnBr_3$. Reprinted with permission from [34]. Copyright (2023) American Chemical Society.

in water. The resulting precipitates are formed, filtered, and washed. Finally, the particles are annealed at a high temperature to form the final product.

Figure 6.8 represents a stepwise coprecipitation strategy for the synthesis of a stable $CsSnBr_3$ perovskite by a facile and low-cost strategy at room temperature [34]. Herein, ethanol as the solvent and salicylic acid as the additive were proposed for the synthesis of a $CsSnBr_3$ perovskite with a highly stable cubic phase. The results showed that the use of the ethanol solvent and the salicylic acid additive can not only effectively prevent the oxidation of Sn^{2+} during the synthesis processes but also stabilize the synthesized $CsSnBr_3$ perovskite. The absorption properties of the perovskite film remained consistent, and its photoluminescent (PL) intensity was significantly preserved at around 69% even after being stored for ten days. This performance was superior to that of a bulk $CsSnBr_3$ perovskite film synthesized using the spin-coating method, which experienced a decrease in PL intensity to 43% after only 12 h of storage. Clearly, this method is an efficient route for the synthesis of perovskite materials at room temperature.

6.3.2 Biotemplate-supported synthesis

The family of microorganisms provides various biological templates that can be used as an inspiration for designing and synthesizing various perovskite materials with different micro- and nanostructures and enhanced specific surface area. Different biological samples, such as algae, bacteria, fungi, viruses, etc. can be used, which offer the benefits of fast growth with control over the material's shape, size, and structure. In addition, biological materials, such as plant leaves, microorganisms, etc. provide good templates for the development of materials, and the unique 3D designs of these samples can be mimicked in the desired materials [26].

Figure 6.9 depicts the introduction of a small quantity of the precious metal Pt into $LaNiO_3$, leading to the formation of $LaNi_{1-x}Pt_xO_3$ with a perovskite structure. In this process, Pt is dispersed atom by atom within the crystal lattice of the perovskite and is supported on SiO_2 which has a high specific surface area. After reduction, the resulting product is $Pt/LaNiO_3/SiO_2$. This approach enables Pt to be incorporated into the perovskite lattice, resulting in a uniform dispersion of Pt atoms at the atomic level. In addition, there is a strong interaction between the Pt atoms and the perovskite material, which effectively prevents the sintering of the Pt atoms [35].

The advantages and disadvantages of the methods used to synthesize perovskite materials, including physical and chemical methods, are listed in table 6.2.

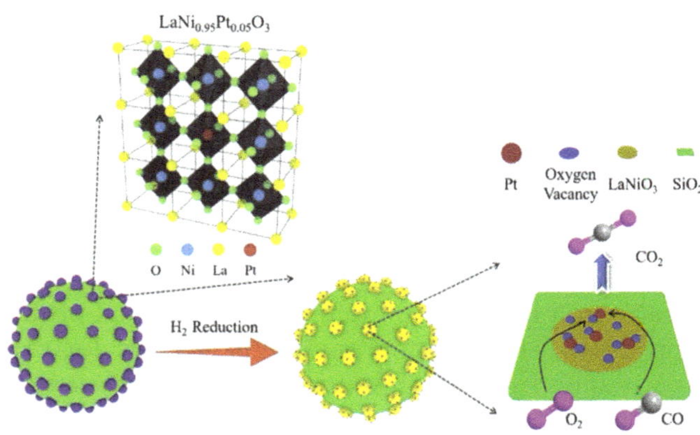

Figure 6.9. A schematic illustration of a process used to load highly dispersed Pt on $LaNiO_3$ which is on the surface of SiO_2 by reducing $LaNi_{1-x}Pt_xO_3/SiO_2$. Reprinted from [35], Copyright (2020), with permission from Elsevier.

Table 6.2. A summary of the advantages and disadvantages of common perovskite synthesis methods. Reprinted from [36], Copyright (2022), with permission from Elsevier.

Synthesis methods	Synthesis sub methods	Advantages	Disadvantages
Physical	High-temperature treatment	• Highly scalable • Simple	• Irregular and nonuniform morphology • Difficult to obtain single-phase and complex perovskites • Generally low specific surface area • High energy consumption
	• Mechanochemical • Reaction	• Highly scalable • Simple	• Difficult to control the morphology • Difficult to obtain single-phase and complex perovskites • High energy consumption

(Continued)

Table 6.2. (*Continued*)

Synthesis methods	Synthesis sub methods	Advantages	Disadvantages
	Radiation	• Short crystallization time • Produces uniform perovskite	• High energy consumption • Requires specialized instrument
Chemical	Coprecipitation	• Highly scalable • Simple	• Generally low crystallinity without additional post-treatment
	Sol–gel	• Highly scalable • Homogeneous and tunable morphology • Mild conditions	• Complex synthesis route • May involve toxic chemicals
	Hydrothermal/ solvothermal	• Homogeneous and tunable morphology	• Requires high temperature and pressure • May involve toxic chemicals

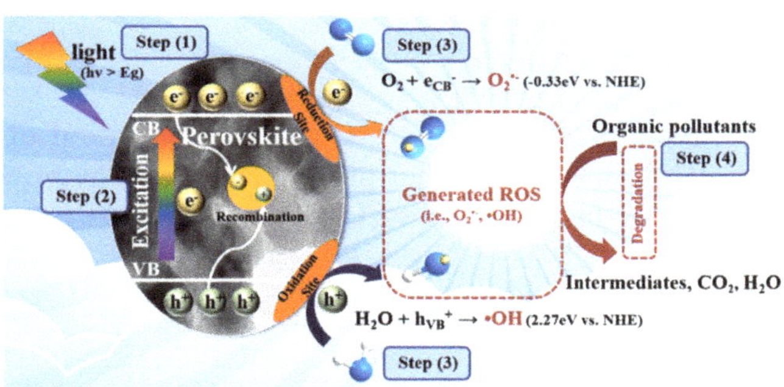

Figure 6.10. The mechanism of ROS production and pollutant degradation performed by perovskite photocatalysts. Reprinted from [10], Copyright (2021), with permission from Elsevier.

6.4 Perovskite photocatalysts for organic degradation

Perovskite-based photocatalytic agents are considered to be good candidates for the effective removal of organic pollutants from wastewater. Recently, many studies have focused on improving their strengths, enhancing the stability of perovskites, improving quantum efficiency, etc [37–40]. Figure 6.10 displays the photocatalytic processes of perovskites, in four steps: (i) light energy interacts with the perovskite material with the requirement that the energy of irradiation must be greater than E_g of the semiconductor. (ii) This activates electrons in the VB region of the semiconductor, causing them to jump to the CB. This process leaves behind positive charges in the VB, which are called holes (h^+). These charges are key factors in carrying out the oxidation (h^+) and reduction reactions (e^-) in photocatalysis. (iii) Complex chemical reactions occur, creating components with strong reducing and oxidizing properties, such as $^{\bullet}OH$ and $^{\bullet}O_2^-$. (iv) Reactive radicals degrade the organic pollutants to form the intermediates CO_2 and H_2O (step 4).

The photocatalytic activities of selected perovskite-based and P25 photocatalysts for the degradation of organic dyes are listed in table 6.3.

6.4.1 Photocatalysis

As mentioned above, materials made from perovskites can absorb solar energy to create oxidizing sites (electron acceptors) and reducing areas (electron donors) for the purpose of oxidating organic pollutants into CO_2, H_2O, and other oxoanions. In photocatalytic degradation, photogenerated holes (h^+) can directly oxidize organic compounds into various intermediates or CO_2 and H_2O. Generally, OH^- derived from adsorbed H_2O is oxidized to hydroxyl radicals ($OH^{\bullet}$) by transferring h^+ from organic compounds to CO_2 and H_2O. Equations (6.3)–(6.5) describe the photocatalytic degradation reactions of organic compounds:

$$\text{Organic} + h^+ \rightarrow \text{intermediates}/CO_2 + H_2O \tag{6.3}$$

$$h^+ + OH^- \rightarrow OH^{\bullet} \tag{6.4}$$

$$\text{Organic} + OH^{\bullet} \rightarrow CO_2 + H_2O \tag{6.5}$$

However, the introduction of O_2 can inhibit the recombination of h^+ and e^- by forming $^{\bullet}O_2^-$ radicals, which then further protonate to form hydroperoxyl radicals ($^{\bullet}HO_2$) and subsequently hydroperoxide (H_2O_2), as depicted in equations (6.6)–(6.9):

$$(O_2)_{ads} + e^- \rightarrow {}^{\bullet}O_2^- \tag{6.6}$$

$$O_2 + OH \rightarrow HOO \tag{6.7}$$

$$HOO + e^- \rightarrow {}^{\bullet}HO_2^- \tag{6.8}$$

$${}^{\bullet}HO_2^- + H^+ \rightarrow H_2O_2 \tag{6.9}$$

Figure 6.11 illustrates the use of a hybrid material based on a lead-free halide perovskite which is activated by light. This hybrid material exhibits enhanced photoelectrical effects and utilizes synergistic surface activation mechanisms. As a result, it supports an efficient process for converting aromatic C(sp3)–H bonds under visible light and in the presence of air.

Table 6.3. The photocatalytic activity of selected perovskite-based and P25 photocatalysts for the degradation of organic dyes [41].

Photocatalyst	BG (eV)	Dye type and/or concentration	Degradation rate ($1 - C/C_0$, %)	Catalyst loading	Light type	Preparation method
P25	3.2	RB, 20 μmol l^{-1}	3	5 g l^{-1}	$\lambda > 400$ nm	Commercial
$LaFeO_3$(500)			24			Sol–gel
$LaFeO_3$(600)			16			
$LaFeO_3$(700)			13			
$LaFeO_3$(800)			9			
$LaFeO_3$ nanocubes	2.01	RB	76.81 at 180 min	1 g l^{-1}	$\lambda > 400$ nm	Hydrothermal
$LaFeO_3$ nanorods	2.05		88.36 at 180 min			
$LaFeO_3$ nanospheres	2.1		90.8 at 180 min			
P25	3.2		5.1 at 180 min			Commercial
$LaNiO_3$	2.26	MO, 10 mg l^{-1}	74.9 at 5h	2 g l^{-1}	$\lambda > 400$ nm	Sol–gel
$CeCo_{0.05}Ti_{0.95}O_{3.97}$	1.57	Nile blue, 30 ppm	91 at 3h	3 g l^{-1}	254–310 and 410–500 nm	Sol–gel
$NaTaO_3$	4.13	MB, 2×10^{-4} mol l^{-1}	60 at 200 min	50 mg	$\lambda > 420$ nm	Hydrothermal
2.47 mol% Cr-doped $NaTaO_3$	3.95		70 at 200 min			
6.35 mol% Cr-doped $NaTaO_3$	3.28		~35 at 200 min			
P25	3.2	RB, 5×10^{-6} mol l^{-1}	75 at 150 min	2 g l^{-1}	$\lambda > 400$ nm	Commercial
$K_2La_2Ti_3O_{10}$	3.63		10 at 150 min			Steady-state reaction (SSR)
N-$K_2La_2Ti_3O_{10}$	3.59		20 at 150 min			SSR + NH_3 gas
CN-$K_2La_2Ti_3O_{10}$	2.92		90 at 150 min			SSR + urea
P25	3.2	MO, 0.02 g l^{-1}	~6 at 30 min	0.8 g l^{-1}	$\lambda > 400$ nm	Commercial
$K_2La_2Ti_3O_{10}$	3.69		~2 at 30 min			SSR
N-$K_2La_2Ti_3O_{10}$	3.44		~30 at 30 min			SSR + NH_3 gas
$LaFeO_3$		MB, 10 mg l^{-1}	~50 at 90 min	0.5 g l^{-1}	$\lambda > 400$ nm	Reverse microemulsion
$La_{0.9}Ca_{0.1}FeO_3$			~80 at 90 min			

$La_2Ti_2O_7$		MO, 1×10^{-5} mol l^{-1}	~55 at 90 min	1 g l^{-1}	UV, $\lambda = 254$ nm	Polymeric complex
$La_{1.5}Pr_{0.5}Ti_2O_7$			~25 at 90 min			
$La_{1.5}Gd_{0.5}Ti_2O_7$			90 at 90 min			
$La_{1.5}Er_{0.5}Ti_2O_7$			~50 at 90 min			
$LaFeO_3$	2.36	MB, 10 mg l^{-1}	~100 at 90 min	2 g l^{-1}	$\lambda > 400$ nm	Microwave-assisted route
P25	3.2		20 at 90 min			Commercial
$LaFeO_3$	2.1	RB	490 at 180 min	1 g l^{-1}	$\lambda > 400$ nm	Hydrothermal
P25	3.2	MO, 10 mg l^{-1}	~85 at 100 min	0.2 g l^{-1}	UV	Commercial
$LaCoO_3$	2.07		~85 at 100 min			Surface-ion adsorption method
P25	3.2	MB, 10 mg l^{-1}	~85 at 100 min			Commercial
$LaCoO_3$	2.07		~85 at 100 min			Surface-ion adsorption method
KNb_3O_8	3.06	Acid red G, 50 mg l^{-1}	63.03 at 60 min	1 g l^{-1}	UV, $\lambda = 253.7$ nm	SSR
0.3 wt% Cu-doped KNb_3O_8			93.23 at 60 min			
2 wt% Cu-doped KNb_3O_8			83.92 at 60 min			

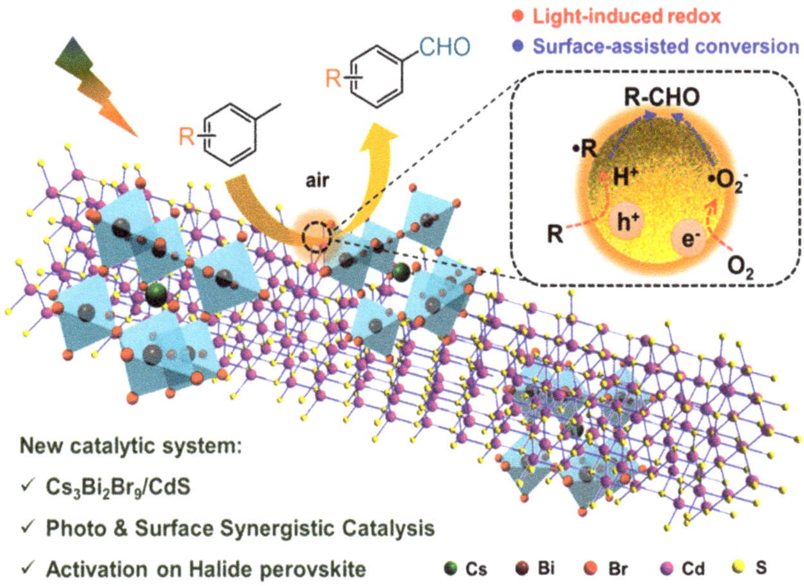

Figure 6.11. A schematic illustration of the photocatalytic conversion of aromatic C(sp3)–H bonds by a $Cs_3Bi_2Br_9$/CdS hybrid photocatalyst under air and visible light. The conversion involves the synergistic activation of photo and surface catalysis. Reprinted from [42], Copyright (2023), with permission from Elsevier.

6.4.2 Fenton and photo-Fenton catalysis

The Fenton and photo-Fenton processes are both AOPs that involve the generation of radicals for oxidation. The distinction between them lies in the origin of these radicals, as shown in figure 6.12. Therein, hydroxyl radicals ($^{\bullet}OH$) can be generated through the catalytic decomposition of hydrogen peroxide (H_2O_2) by ferric ions in the Fenton reaction. Meanwhile, the photo-Fenton reactions occur in the presence of UV or visible light, and the oxidative radicals derived from Fe^{3+}/H_2O_2 are produced through the action of the light source. In photocatalytic oxidation processes, electron–hole pairs are generated within the catalyst upon exposure to UV light, and the oxidative radicals are formed at the interface between the catalyst and water [43].

The Fenton method has attracted great interest due to its convenience and effectiveness. Notably, the Fenton method can produce many hydroxyl radicals ($^{\bullet}OH$) by introducing a divalent iron solution and hydrogen peroxide [44], as shown in equation (6.10).

$$Fe^{2+} + H_2O_2 \rightarrow Fe^{3+} + OH^- + OH^{\bullet} \tag{6.10}$$

The main result of the Fenton reaction, which is primarily governed by equation (6.10), is the formation of hydroxyl radicals ($HO^{\bullet}$). However, in the photo-Fenton process, two additional reactions occur simultaneously [45, 46]. First, there is the photolysis of hydrogen peroxide (H_2O_2) through the absorption of light energy ($h\nu$),

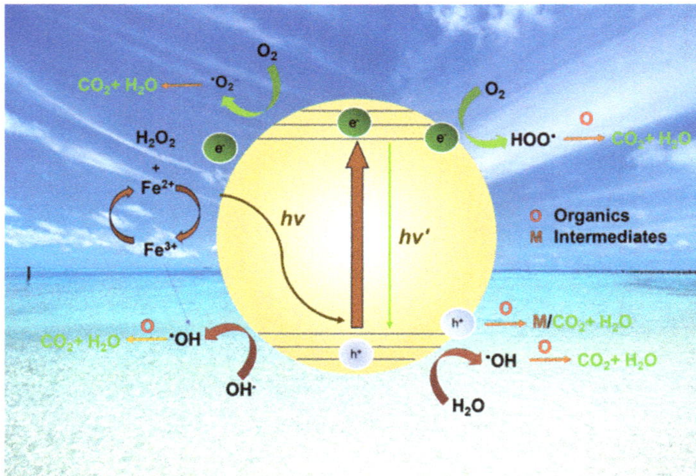

Figure 6.12. A schematic illustration of the photo-Fenton reactions of perovskites.

leading to the formation of two hydroxyl radicals ($HO^{\bullet}$) as shown in equation (6.11). This photolytic reaction occurs when H_2O_2 absorbs photons and undergoes a cleavage reaction, resulting in the generation of highly reactive hydroxyl radicals.

$$H_2O_2 \overset{hv}{\rightarrow} 2\ OH^{\bullet} \tag{6.11}$$

Second, there is the photoreduction of ferric ions (Fe^{3+}) in the presence of water (H_2O) and light energy ($h\nu$). This reaction, described by equation (6.11), involves the absorption of photons by Fe^{3+}, leading to the reduction of Fe^{3+} to ferrous ions (Fe^{2+}), the generation of hydroxyl radicals ($HO^{\bullet}$), and the release of a hydrogen ion (H^{+}). This photoreduction process contributes to the overall generation of hydroxyl radicals in the photo-Fenton process.

$$Fe^{3+} + H_2O \quad hv \quad Fe^{2+} + OH^{\bullet} + H^{+} \tag{6.12}$$

Together, the Fenton reaction (equation (6.10)), the photolysis of H_2O_2 (equation (6.18)), and the photoreduction of Fe^{3+} (equation (6.11)) result in the production of hydroxyl radicals ($HO^{\bullet}$) in the photo-Fenton process. These highly reactive radicals play a crucial role in the degradation of organic pollutants by oxidizing and breaking down their chemical bonds.

6.4.3 PMS activation

PMS is frequently employed as an oxidizing agent in AOPs to generate sulfate radicals. These radicals can be activated through various methods. Instead of relying on energy-based activation methods such as solar/UV radiation, ultrasound, electricity, or heat, catalytic activation is more commonly employed. This is because it offers advantages such as lower cost and higher efficiency in decomposing PMS using transition-metal ions and heterogeneous catalysts. In AOP systems, the

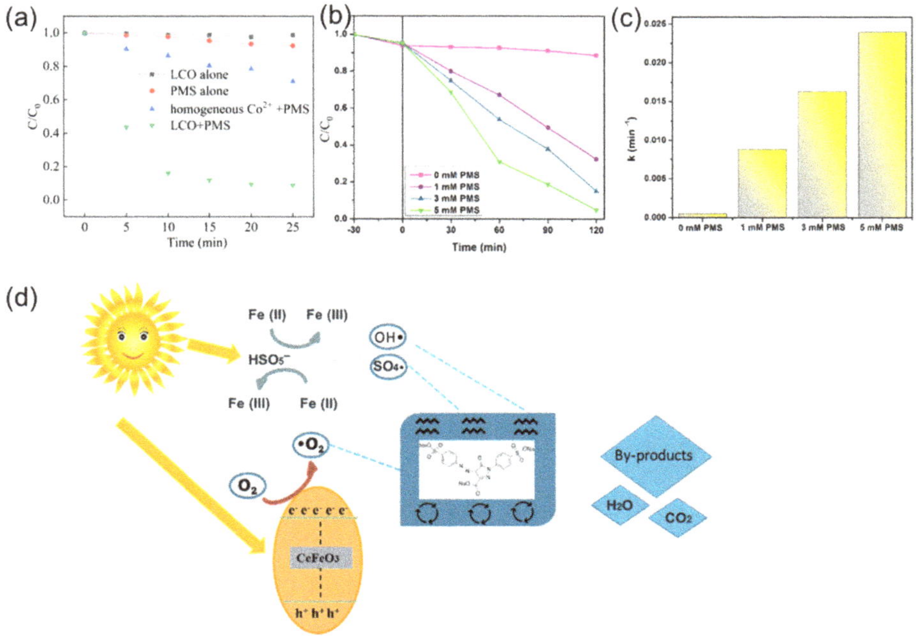

Figure 6.13. (a) BPA degradation in different reaction systems. Reaction conditions: [BPA] = 0.05 mM, [PMS] = 1.0 mM, [LCO] = 0.5 g l^{-1}, initial pH = 6.8. Reproduced from [47] with permission from the Royal Society of Chemistry. (b–c) Photocatalytic degradation curves of $CeFeO_3$ samples in the PMS/Vis system. (d) The proposed mechanism of the photoactivation of PMS over $CeFeO_3$ samples. Reprinted from [48], Copyright (2023), with permission from Elsevier.

presence of heterogeneous perovskite catalysts allows the generation of numerous ROS, including $HO^{\bullet}$ radicals, singlet oxygen, and ${}^{\bullet}O_2^-$. The activation of PMS by heterogeneous catalysts has gained significant attention because it enables easier separation of the catalysts, which helps to prevent secondary pollution issues that may arise in homogeneous systems.

Zhong *et al* synthesized La_2CoO_4+δ (LCO) perovskite and active PMS to degrade bisphenol A (BPA) [47]. As shown in figure 6.13a, LCO+PMS had a highly efficient photocatalytic performance, which was superior to those of LCO alone, PMS, or homogeneous Co^{2+}+PMS systems. Based on figures 6.13(b) and (c), it can be observed that the activity of bare PMS was relatively low, achieving only 12.6% degradation. This suggests that the activation of PMS on the surface of $CeFeO_3$ plays a more prominent role compared to light irradiation, which is illustrated in figure 6.13(c). Two primary interactions involving persulfate ions can be credited for the heightened activity observed in $CeFeO_3$/PMS/Vis systems. First, visible light can activate PMS, leading to its activation and the generation of reactive species. Second, the presence of ferrite sites within $CeFeO_3$ aids in the activation of PMS, facilitating the desired reactions. These interactions lead to the generation of free radicals, specifically sulfate radicals (${}^{\bullet}SO_4^-$) and hydroxyl radicals

($HO^•$), which are responsible for the decomposition of dye molecules. The proposed overall activation reactions are summarized in equations (6.13)–(6.15):

$$HSO_5^- + h\nu \rightarrow SO_4^• + OH^• \tag{6.13}$$

$$HSO_5^- + Fe(III) \rightarrow SO_5^• + H^+ + Fe(II) \tag{6.14}$$

$$HSO_5^- + Fe(II) \rightarrow SO_4^• + OH^• + Fe(III) \tag{6.15}$$

Figure 6.13(c) shows that the visible-light photons are primarily absorbed by the surface of $CeFeO_3$ in the $CeFeO_3$/PMS/vis system. This absorption leads to the generation of photoexcited electrons in the conduction band and holes in the valence band of $CeFeO_3$. This $CeFeO_3$ can react with adsorbed oxygen molecules, leading to the formation of $^•O_2^-$ radicals through a process known as electron transfer. These $^•O_2^-$ radicals are highly reactive and can participate in subsequent oxidation reactions.

Simultaneously, PMS is also irradiated with light, resulting in the formation of sulfate radicals ($^•SO_4^-$) and hydroxyl radicals ($^•OH$). These radicals are generated through the activation of PMS by the absorbed light energy. The presence of PMS contributes to the overall radical generation in the system. In addition, the Fe^{3+} state within the $CeFeO_3$ perovskite structure can undergo oxidation, leading to the formation of additional radicals. This further enhances the photocatalytic oxidation processes occurring in the system.

Clearly, the combined action of $^•O_2^-$ radicals, sulfate radicals, hydroxyl radicals, and other reactive species generated within the $CeFeO_3$/PMS/vis system contributes to the overall positive effect of the radicals. These radicals participate in various oxidation reactions, effectively degrading organic pollutants, such as dye molecules. Overall, the interplay between photoexcited electrons in $CeFeO_3$, $^•O_2^-$ radicals and PMS activation through light irradiation results in the generation of a variety of radicals that drive the photocatalytic oxidation processes within the system, leading to the enhanced degradation of pollutants.

6.5 Conclusions

Perovskite materials possess exceptional crystallinity and unique features, making them highly promising as semiconductor photocatalysts for the degradation of organic pollutants. The ability to fine-tune their physicochemical properties by adjusting their chemical composition offers ample opportunities for developing novel nanocomposites. However, pristine perovskites suffer from limitations, i.e. low photocatalytic efficiency, poor stability, inadequate utilization of solar energy, rapid electron–hole recombination, and low redox potential. To overcome these challenges, researchers have capitalized on the flexibility of regulating perovskite properties to develop efficient photocatalysts for water remediation. Numerous perovskite-based nanocomposites have been synthesized and studied using various methods and strategies, including cationic substitution with dopants, downsizing and morphological modifications, and coupling with AOPs. While designing and preparing novel perovskite-based nanocomposites, particularly through partial and full cationic substitution, is an elegant approach, it remains a challenging task.

The photocatalytic activity of perovskites is improved by controlling the bandgap, defects, morphology, etc. In addition, the photo-Fenton method and PMS activation by perovskite-based nanomaterials were discussed in this chapter. Perovskite-based nanomaterials have demonstrated encouraging outcomes as activators for PMS in applications. They have exhibited promising results in the removal of diverse organic pollutants, including endocrine disruptors, antibiotics, and dyes However, it is worth noting that metal halide perovskites face challenges related to their stability, especially in the presence of moisture and heat. To achieve broader success in practical applications utilizing perovskite-based nanomaterials, the following research directions have been identified:

(a) developing new perovskites and coupling perovskites with other semiconductors or metal nanoparticles;
(b) investigating the photocatalysis of realistic wastewater samples;
(c) upscaling perovskite catalysts;
(d) improving their stability and long-term performance to make them commercially viable;
(e) developing environmentally friendly synthesis methods;
(f) decorating 3D-printed materials with perovskite photocatalysts for use in industrial production, and
(g) improving catalyst separation (membrane filtration is commonly used for catalyst separation but using membranes for perovskite-based catalyst recovery can be costly and complex due to membrane fouling).

References

[1] Lomonaco T *et al* 2020 Release of harmful volatile organic compounds (VOCs) from photo-degraded plastic debris: a neglected source of environmental pollution *J. Hazard. Mater.* **394** 122596

[2] Chaudhry G R and Chapalamadugu S J 1991 Biodegradation of halogenated organic compounds *Microbiol Rev.* **55** 59–79

[3] Singh S, Yadav R, Sharma S and Singh A N 2023 Arsenic contamination in the food chain: a threat to food security and human health *J. Appl. Biol. Biotechnol.* **11** 24–33

[4] Carpenter D O 2013 *Effects of Persistent and Bioactive Organic Pollutants on Human Health* (New York, NY: Wiley) DOI:10.1002/9781118679654

[5] Balk S J, Carpenter D O and Corra L *et al* 2010 *Persistent Organic Pollutants: Impact on Child Health* 9789241501101 World Health Organization Technical document

[6] Fuller R *et al* 2022 Pollution and health: a progress update *Lancet Planet. Health* **6** e535–47 (Corrected 18 May 2022 at https://doi.org/10.1016/S2542-5196(22)00145-0)

[7] Ahmed S *et al* 2011 Advances in heterogeneous photocatalytic degradation of phenols and dyes in wastewater: a review *Water, Air, Soil Pollut.* **215** 3–29

[8] Huang H, Pradhan B, Hofkens J, Roeffaers M B and Steele J A 2020 Solar-driven metal halide perovskite photocatalysis: design, stability, and performance *ACS Energy Lett.* **5** 1107–23

[9] Chen Z-Y, Huang N-Y and Xu Q 2023 Metal halide perovskite materials in photocatalysis: design strategies and applications *Coord. Chem. Rev.* **481** 215031

[10] Wei K, Faraj Y, Yao G, Xie R and Lai B 2021 Strategies for improving perovskite photocatalysts reactivity for organic pollutants degradation: a review on recent progress *Chem. Eng. J.* **414** 128783

[11] Eames C *et al* 2015 Ionic transport in hybrid lead iodide perovskite solar cells *Nat. Commun.* **6** 7497

[12] Kuru T *et al* 2023 Photodeposition of molybdenum sulfide on $MTiO_3$ (M: Ba, Sr) perovskites for photocatalytic hydrogen evolution *J. Photochem. Photobiol. A: Chem.* **436** 114375

[13] Luo J *et al* 2021 Halide perovskite composites for photocatalysis: a mini review *EcoMat* **3** e12079

[14] Kong J, Yang T, Rui Z and Ji H 2019 Perovskite-based photocatalysts for organic contaminants removal: current status and future perspectives *Catal. Today* **327** 47–63

[15] Tao S *et al* 2019 Absolute energy level positions in tin- and lead-based halide perovskites *Nat. Commun.* **10** 2560

[16] Deschler F *et al* 2014 High photoluminescence efficiency and optically pumped lasing in solution-processed mixed halide perovskite semiconductors *J. Phys. Chem. Lett.* **5** 1421–6

[17] Dey K, Roose B and Stranks S D 2021 Optoelectronic properties of low-bandgap halide perovskites for solar cell applications *Adv. Mater.* **33** 2102300

[18] Prasanna R *et al* 2017 Band gap tuning via lattice contraction and octahedral tilting in perovskite materials for photovoltaics *J. Am. Chem. Soc.* **139** 11117–24

[19] Irshad M *et al* 2022 Photocatalysis and perovskite oxide-based materials: a remedy for a clean and sustainable future *RSC Adv.* **12** 7009–39

[20] Liu W, Lee J-S and Talapin D V 2013 III–V nanocrystals capped with molecular metal chalcogenide ligands: high electron mobility and ambipolar photoresponse *J. Am. Chem. Soc.* **135** 1349–57

[21] Lim J *et al* 2022 Long-range charge carrier mobility in metal halide perovskite thin-films and single crystals via transient photo-conductivity *Nat. Commun.* **13** 4201

[22] Hu H *et al* 2020 Nucleation and crystal growth control for scalable solution-processed organic–inorganic hybrid perovskite solar cells *J. Mater. Chem.* A*8 1578–603*

[23] Steirer K X *et al* 2016 Defect tolerance in methylammonium lead triiodide perovskite *ACS Energy Lett.* **1** 360–6

[24] Farhad F T *et al* 2018 $MAPbI_3$ and $FAPbI_3$ perovskites as solar cells: Case study on structural, electrical and optical properties *Results Phys.* **10** 616–27

[25] Hosseinian Ahangharnejhad R *et al* 2021 Protecting perovskite solar cells against moisture-induced degradation with sputtered inorganic barrier layers *ACS Appl. Energy Mater.* **4** 7571–8

[26] Kumar A, Kumar A and Krishnan V J A c 2020 Perovskite oxide based materials for energy and environment-oriented photocatalysis *ACS Catal.* **10** 10253–315

[27] Salavati-Niasari M *et al* 2016 Synthesis, characterization, and morphological control of $ZnTiO_3$ nanoparticles through sol–gel processes and its photocatalyst application *Adv. Powder Technol.* **27** 2066–75

[28] Parida K, Reddy K, Martha S, Das D and Biswal N 2010 Fabrication of nanocrystalline $LaFeO_3$: an efficient sol–gel auto-combustion assisted visible light responsive photocatalyst for water decomposition *Int. J. Hydrog. Energy* **35** 12161–8

[29] Wang S *et al* 2018 Sol-gel preparation of perovskite oxides using ethylene glycol and alcohol mixture as complexant and its catalytic performances for CO oxidation *ChemistrySelect* **3** 12250–7

[30] Zhao H, Duan Y and Sun X 2013 Synthesis and characterization of $CaTiO_3$ particles with controlled shape and size *New J. Chem.* **37** 986–991,

[31] Athayde D D *et al* 2016 Review of perovskite ceramic synthesis and membrane preparation methods *Ceram. Int.* **42** 6555–71

[32] Moshtaghi S, Gholamrezaei S and Niasari M S 2017 Nano cube of $CaSnO_3$: facile and green co-precipitation synthesis, characterization and photocatalytic degradation of dye *J. Mol. Struct.* **1134** 511–9

[33] Junwu Z *et al* 2007 Solution-phase synthesis and characterization of perovskite LaCoO3 nanocrystals via a co-precipitation route *J. Rare Earths* **25** 601–4

[34] Cao L *et al* 2023 Highly ambient stable $CsSnBr_3$ perovskite via a new facile room-temperature 'Coprecipitation' strategy *ACS Appl. Mater. Interfaces* **15** 30409–16

[35] Zhang S *et al* 2020 SiO2 supported highly dispersed Pt atoms on $LaNiO_3$ by reducing a perovskite-type oxide as the precursor and used for CO oxidation *Catal. Today* **355** 222–30

[36] Koo P-L, Jaafar N F, Yap P-S and Oh W-D 2022 A review on the application of perovskite as peroxymonosulfate activator for organic pollutants removal *J. Environ. Chem. Eng.* **10** 107093

[37] Peng Y, Albero J and Garcia H 2019 Surface silylation of hybrid benzidinium lead perovskite and its influence on the photocatalytic activity *ChemCatChem* **11** 6384–90

[38] Schanze K S, Kamat P V, Yang P and Bisquert J 2020 Progress in perovskite photocatalysis *ACS Energy Lett.* **5** 2602–4

[39] Li Q and Lian T 2019 Ultrafast charge separation in two-dimensional CsPbBr3 perovskite nanoplatelets *J. Phys. Chem. Lett.* **10** 566–73

[40] DuBose J T and Kamat P V 2022 Efficacy of perovskite photocatalysis: challenges to overcome *ACS Energy Lett.* **7** 1994–2011

[41] Wang W, Tadé M O and Shao Z 2015 Research progress of perovskite materials in photocatalysis- and photovoltaics-related energy conversion and environmental treatment *Chem. Soc. Rev.* **44** 5371–408

[42] Yang Y *et al* 2023 Synergistic surface activation during photocatalysis on perovskite derivative sites in heterojunction *Appl. Catal.* B **323** 122146

[43] Orak C, Atalay S, Ersöz G J S S and Technology 2017 Photocatalytic and photo-Fenton-like degradation of methylparaben on monolith-supported perovskite-type catalysts *Sep. Sci. Technol.* **52** 1310–20

[44] Huang C-W *et al* 2022 Solar-light-driven $LaFe_xNi_{1-x}O_3$ perovskite oxides for photocatalytic Fenton-like reaction to degrade organic pollutants *Beilstein J. Nanotechnol.* **13** 882–95

[45] Rojas-Cervantes M L and Castillejos E 2019 Perovskites as catalysts in advanced oxidation processes for wastewater treatment *Catalysts* **9** 230

[46] Shinichi E *et al* 2014 Fenton chemistry at aqueous interfaces *PNAS* **111** 623628

[47] Zhong X, Wu W, Jie H and Jiang F 2023 $La_2CoO_{4+\delta}$ perovskite-mediated peroxymonosulfate activation for the efficient degradation of bisphenol A *RSC Adv.* **13** 3193–203

[48] Tuna Ö and Bilgin Simsek E 2023 Promoted peroxymonosulfate activation into ferrite sites over perovskite for sunset yellow degradation: optimization parameters by response surface methodology *Opt. Mater.* **142** 114122

Chapter 7

Emerging trends and future prospects in photocatalysis-based environmental remediation and hydrogen production

İbrahim Hakkı Karakaş and Zeynep Karcıoğlu Karakaş

Photocatalysis-based environmental remediation and water splitting technologies are critical to address urgent global challenges such as pollution, environmental degradation, and the need for sustainable energy sources. These technologies have the potential to provide sustainable solutions in an era of increasing emphasis on cleaner production technologies and a growing preference for more environmentally responsible approaches to environmental remediation and energy production processes. In this chapter, we investigate recent advances in photocatalysis-based environmental remediation and hydrogen production technologies and provide insights into the challenges faced in applying the processes or innovations that will likely emerge. Moreover, this chapter sheds light on issues such as enhancing catalytic efficiency, recent advances in the design of new photocatalytic materials, and expanding their application areas. Furthermore, a comprehensive overview of current emerging trends and challenges in photocatalysis and future prospects is presented, and the multifaceted role of photocatalysis technology in environmental remediation and sustainable hydrogen production is explained.

7.1 Introduction

7.1.1 Emerging trends and future prospects in photocatalysis

Photocatalysis-based environmental remediation technologies are a promising and rapidly developing research field with many exciting developments and potential future directions. Photocatalysis is a process based on the use of light-activated catalysts to accelerate chemical reactions. Photocatalytic processes often remove pollutants from the air or water or produce H_2 via water splitting. It is expected that

doi:10.1088/978-0-7503-5697-8ch7

photocatalysis-based environmental remediation applications will gain more attention in the future, as some of the existing limitations will be overcome [1].

Nowadays, many researchers are investigating the development of new photocatalytic materials with improved efficiency, stability, and high selectivity. These studies focus on the photocatalytic use of new materials such as graphene-based catalysts, metal–organic frameworks (MOFs), MXene-based materials, and perovskites. These materials can be used directly as photocatalysts or in combination with other materials [2].

Nanotechnology plays a vital role in photocatalytic processes. When nanoparticles and nanocomposites are used as photocatalysts, the high surface areas of these materials have the potential to significantly increase catalytic activity. Photocatalytic processes can be applied for various environmental purposes, such as air purification, water treatment, water splitting, and the remediation of contaminated soil. Using these processes, it is possible to degrade a wide range of pollutants, such as organic dyes, pharmaceutics, pesticides, and volatile organic compounds (VOCs). New studies of the usability of photocatalytic processes in removing different contaminants continue to be carried out daily. It is expected that the types of pollutants that photocatalytic processes can remove will soon increase, and thus their application areas will become much more diversified. Today, photocatalytic processes can be used for water treatments, the removal of organic and inorganic pollutants, and photocatalytic disinfection [3].

Despite current challenges, research and development efforts are ongoing to address potential issues and make large-scale photocatalytic water splitting a reality. Innovative reactor designs, advanced catalyst materials, improved mass transfer techniques, and a better understanding of system dynamics are some focal areas that are being used to overcome these challenges. As technology advances, the scalability and efficiency of photocatalytic water splitting processes are expected to increase, and these processes are expected to become increasingly important in the renewable energy and hydrogen production sectors.

Photocatalytic processes have also been used in air quality improvement studies to improve indoor and outdoor air quality by removing some organic or inorganic pollutants from the air, including VOCs and nitrogen oxides (NOx). Some of these studies have also developed into commercial products such as self-cleaning coatings. Many new commercial products with similar properties are imminently expected to emerge. As the technology in this field matures and becomes more cost-effective, its adoption for commercial purposes is expected to increase further.

On the other hand, there is also growing interest in integrating photocatalytic processes with renewable energy sources such as solar and wind energy to improve energy efficiency and sustainability. Integrating photocatalytic processes with renewable energy requires careful planning and coordination [4].

Increasing the efficiency of photocatalysis under visible light is an important focus. Since visible light is more abundant than UV light, this would allow for more efficient solar energy utilization. The use of natural daylight instead of artificial light sources in photocatalytic processes has both advantages and some limitations [5]. Natural daylight is a sustainable and easily accessible energy source.

In photocatalytic processes, it helps to lower operating costs and environmental impact by reducing the need for electricity or artificial light sources. The use of renewable energy sources makes photocatalytic processes compatible with clean energy and sustainability goals. This provides a significant advantage for the usability of photocatalytic processes, especially in large-scale applications. Moreover, the use of natural daylight in photocatalysis can reduce the carbon footprint by reducing energy consumption and thus minimizing the greenhouse gas emissions caused by energy production.

Furthermore, natural daylight has the potential to provide higher energy efficiency, as it contains a broader spectrum of light than the UV–vis light sources currently used as light sources for photocatalytic reactions. In addition to the advantages of using daylight in photocatalytic processes, there are also some limiting challenges. Natural daylight varies in intensity and availability due to the weather conditions, time of day, and geographical location. This dictates that daylight-driven photocatalytic processes can only be actively applied when daylight is available. Therefore, these variations can significantly affect the consistency and predictability of photocatalytic reactions. Moreover, the spectrum and intensity of natural sunlight vary with the geographical location, time of year, and time of day. These differences can negatively affect the performance of photocatalytic materials optimized for specific wavelengths of light. Energy storage systems such as batteries may be required to sustain continuous photocatalytic reactions in the absence of daylight. These systems not only make construction more complex but also significantly increase costs [6–8].

Moreover, some photocatalytic reactions may require light sources of exceptionally high intensity. It may not be possible to provide this irradiance using natural daylight alone. Thus, for such applications, it may be necessary to use additional artificial light sources. The choice between natural daylight and artificial light sources in photocatalysis depends on the application and requirements. For example, natural sunlight can be used for water purification or improving air quality in remote or off-grid areas. In contrast, artificial light sources should be used for precise and consistent reactions in controlled environments.

In practice, many photocatalytic applications aim for an optimal mix of natural daylight and artificial lighting. In these studies, the system utilizes sunlight for photocatalysis and is supplemented with artificial light sources when needed. This approach combines the advantages of sustainability and cost savings with the reliability and control offered by artificial light [9].

Photocatalytic water splitting is a technology that has the potential to play an important role in renewable energy production. Photocatalytic water splitting uses direct UV or sunlight to split water molecules into gaseous hydrogen and oxygen. Hydrogen produced by this method can be used as a clean and renewable energy source [10–12].

Researchers worldwide are currently working to improve the efficiency of photocatalytic water splitting systems. These studies often focus on developing new materials for photocatalysts and designing new reactors to capture more of the solar spectrum and convert it into chemical energy [13, 14].

Today, metal oxides such as titanium dioxide (TiO_2), zinc oxide (ZnO), and many different semiconductor materials are used as photocatalysts in photocatalytic water splitting processes. New and more efficient photocatalysts are expected to be developed as a result of advances in materials science in the near future. Moreover, it may be possible to produce composites of some of the currently used photocatalysts with other materials and thus improve the catalytic activity of these materials. In addition, some recent studies have shown that new materials, such as perovskite compounds and quantum dots, can be very effective in photocatalytic water splitting processes. Further research and development efforts in this field will lead to even more significant progress [15–18].

It is also essential to develop new technologies that will make it possible to provide the light required for photocatalytic water splitting processes directly from the Sun. In today's photocatalytic water splitting processes, UV lamps that emit continuous radiation at a specific wavelength are generally preferred as light sources. However, these lamps' energy consumption increases these processes' environmental footprint. Improving the integration of photocatalytic water splitting with solar technologies such as photovoltaic cells and concentrated solar power can make large-scale hydrogen production more efficient, environmentally friendly, and practical [11, 19–21].

7.2 Scalability challenges for photocatalytic applications

Researchers and engineers have been working to design scalable and cost-effective photocatalysis processes for large-scale applications. However, scaling up photocatalytic water splitting processes for industrial applications presents significant challenges [22]. Scaling up photocatalytic water splitting processes can present challenges in transitioning from primarily laboratory-scale experiments to industrial or commercial applications. As the scale increases, ensuring efficient and homogeneous light absorption or distribution across a larger reactor becomes more difficult. Maximizing the utilization of sunlight or artificial light sources requires innovative engineering solutions. Moreover, mass transfer limitations can also be an issue in scaling up processes. Mass transfer limitations become more pronounced in larger reactors. Especially in large-scale heterogeneous catalysis systems, it can be much more challenging to deliver reactants to the catalyst surface and efficiently remove the products (hydrogen and oxygen) from the reactor than in laboratory-scale systems. Another challenge associated with scaling up processes is to ensure that operating parameters such as temperature, pH, and flow rates, which have the potential to directly affect process efficiency, are homogeneous in all regions of the reactor. In addition, in large-scale systems, some parts of the reactor must be sufficiently exposed to light [23–25].

For high-efficiency photocatalytic processes, larger photocatalytic reactors need to be designed and manufactured to ensure that the ideal values of parameters such as optimized temperature, pH, and flow rate remain constant and, at the same time, ensure the homogeneous distribution of light throughout the reactor. For this purpose, new reactors should be designed by considering efficient mixing and flow models [26]. In addition, problems with uniform and homogeneous catalyst distribution throughout the reactor may arise in large reactors. A homogeneous catalyst distribution is fundamental in large-scale systems to achieve stable and highly efficient photocatalytic reactions. If homogeneous catalyst distribution is not achieved, this can lead to variations in reaction rates and a decrease in overall efficiency. To avoid these problems, designing and manufacturing new reactors is essential [27].

Moreover, large quantities of catalysts will be required for large-scale systems. Given that the materials currently used as photocatalysts are nanoscale advanced materials, it can be challenging to produce photocatalytic materials in the quantities needed for large-scale applications. The methods used to synthesize and prepare photocatalysts at the laboratory scale may need to be more easily scalable to meet commercial demands. In this case, scaled up processes for these catalysts may also need to be meticulously designed. In addition, factors such as the stability and durability of the photocatalysts produced may also become an issue during scale-up [28, 29]. Commercial applications often require longer catalyst lifetimes, and problems due to photocatalyst degradation or fouling can become more pronounced in larger systems. Significant progress is expected in the development of reusable photocatalyst materials that have improved chemical and mechanical stability.

Another aspect that needs to be considered concerning photocatalytic water splitting processes is economic viability. The scale-up of photocatalytic water splitting processes must be economically viable. Factors such as the reactor's design, manufacture, and maintenance and the cost of production, stability, and reusability of catalyst materials are directly related to process economics [30]. To assess the economic viability of the process, all the costs incurred need to be evaluated together with the amount of hydrogen produced by this process and any other possible benefits. Moreover, meeting regulatory standards for environmental and safety considerations can require more work in large-scale systems. This is because it is crucial to avoid negative environmental impacts when scaling up photocatalytic processes.

Despite these challenges, much research and development work is underway to overcome these issues and make large-scale photocatalytic water splitting a reality. Scientific studies in this field are focused on designing innovative reactors, developing advanced catalyst materials, and improving mass transfer processes. In the near future, due to increasing environmental concerns and constantly evolving technology, it is predicted that the interest in hydrogen, considered a green energy source, will increase even more. Therefore, it is expected to become increasingly essential to scale up and increase the efficiency of hydrogen production processes by photocatalytic water splitting.

The possible negative environmental impacts of photocatalytic water splitting processes must be addressed further. Currently, researchers are focused on developing methods that use sustainable materials for hydrogen production and minimize the release of pollutants during the hydrogen production process. In light of the data obtained from these studies, it is predicted that more concrete advances will be made, and much cleaner production processes will be designed. In hydrogen technologies, safe hydrogen storage is as necessary as its production. New safe and efficient hydrogen storage technologies should also be developed to store the hydrogen gas produced by photocatalytic water splitting methods in hydrogen fuel cells. Photocatalytic processes are likely to be used in many different commercial-scale applications in the near term. Photocatalytic processes have the potential to overcome various environmental and industrial challenges. Currently, the main factor limiting the availability of photocatalytic methods on a commercial scale is the high cost of these processes [31]. However, shortly, due to advances in materials science, reactor design, and manufacturing techniques, the cost of photocatalytic systems is expected to decrease significantly, making them economically viable for large-scale applications. Some applications in which photocatalytic processes are expected to be applied commercially are summarized below [32, 33]. Photocatalytic processes are likely to play an essential role in water treatment on a commercial scale. They have the potential to simultaneously and efficiently remove many types of contaminants, including organic pollutants, pharmaceuticals, and microorganisms that are likely to be present in water. Photocatalytic processes can simultaneously disinfect water as well as remove contaminants. Therefore, photocatalysis-based treatment technologies can be used in large-scale water treatment systems to improve water quality [33].

Air purification is another area in which photocatalysis-based technologies have been used extensively and successfully. Photocatalysis is well suited for large-scale applications such as indoor air purification in commercial buildings, factories, and public spaces. Coatings and materials with photocatalytic properties have the potential to help reduce the concentration of VOCs and pollutants in both indoor and outdoor air [34–36]. Photocatalysis can also be applied in green hydrogen production through water splitting, which has commercial applications in clean energy and fuel cell technologies. Large-scale hydrogen production plants are predicted to adopt photocatalytic processes as a sustainable and environmentally friendly method of production. Due to photocatalytic technology, it is possible to produce products with self-cleaning surfaces. Self-cleaning surfaces coated with photocatalytic materials, such as glass or building exteriors, are expected to be more widely used in commercial buildings. These materials can significantly reduce the maintenance costs of facilities and improve their esthetics [37].

Thanks to photocatalysis-based technologies, it is possible to remove contaminants from soil areas and groundwater contaminated with various pollutants. Moreover, photocatalytic processes can be applied in agriculture to improve soil quality, remove pesticides, and enhance the decomposition of organic matter. Thus, developing more sustainable and efficient agricultural practices on a commercial scale may be possible. Moreover, photocatalysis-based processes will likely be used for pollutant control and process optimization in industrial processes in various fields such as chemistry, textiles, and food.

In conclusion, the applicability of photocatalysis-based processes in commercial-scale applications in the future will be shaped by the data to be obtained from ongoing research and development efforts. Accordingly, the future of photocatalytic processes depends on the adoption of possible new technologies that will emerge as a result of scientific studies by the relevant industries and the reduction of the costs associated with these processes. The integration of photocatalytic processes into various commercial sectors is expected to expand considerably due to increasing environmental concerns and energy-related challenges.

7.3 Scalability challenges for photocatalytic H_2 production applications

More and more fossil fuels are being consumed every day to meet the ever-increasing energy demand on a global scale. This situation rapidly increases the environmental risks arising from fossil fuels and also leads to the rapid depletion of these resources. Therefore, interest in green or renewable energy sources is growing daily. Hydrogen is seen as a green energy source. However, many current methods for hydrogen production need to be more economically and technically scalable to meet the requirements of large-scale production. Moreover, the methods currently applied for H_2 storage have some significant limitations regarding safety and usefulness. The future of large-scale hydrogen production by photocatalytic oxidation-based methods has great potential. It represents a sustainable and environmentally friendly method for producing hydrogen, which is a clean energy carrier. However, there are several factors to consider and some challenges to overcome when assessing the potential and applicability of these methods for large-scale production [1, 38].

As with other technologies, the scale-up for the large-scale production of H_2 by photocatalytic methods depends on the economic viability of the process. The process must be made cost-effective for photocatalytic hydrogen production to be preferable to other methods currently being applied for hydrogen production.

On the other hand, while hydrogen is considered a clean energy source, the potential environmental impact of the production methods is also significant. Photocatalytic oxidation is considered to be more environmentally friendly, as it does not produce carbon emissions. However, the environmental concerns associated with the production and use of materials must be carefully considered and potential risks minimized.

Scaling up hydrogen production through photocatalytic oxidation for large-scale production is seen as a promising method for creating a sustainable environment. Scaling up H_2 production by photocatalytic methods for large-scale production can only be achieved by overcoming some of the current economic and environmental challenges. Accordingly, advances in materials science, reactor design, and renewable energy technologies are expected to make this method more efficient and cost-effective.

'Scale-up' refers to increasing the capacity or size of a system or production process. In the context of chemical manufacturing, scale-up refers to the transition from a smaller-scale laboratory setup to much larger commercial-scale production. Scale-up is crucial for moving new technologies from the research and development stage into practical applications. There are some key considerations to bear in mind when scaling up for the large-scale production of H_2 production by photocatalytic methods.

Photocatalytic oxidation processes typically use light from a light source to drive the reaction. This light can come from an artificial lighting system or the Sun. However, artificial light sources are not sustainable due to their energy consumption. Accordingly, using sunlight as a light source is an economical and environmentally preferable approach. However, the utilization of sunlight for large-scale H_2 production in photocatalysis-based systems can only be made possible by developing new, more efficient, and cost-effective methods for capturing and converting sunlight. Therefore, with the advances in photovoltaics and the development of better photocatalysts, these processes are soon expected to become more effective and preferable for large-scale H_2 production.

For large-scale H_2 production in photocatalysis-based systems, one of the most critical factors affecting the efficiency of the process is the material used as the photocatalyst. Researchers are continuously working to develop new catalysts that are not only efficient but also cost-effective and readily available and can maintain their stability over long periods. Another important parameter in increasing the yields of reactions in H_2 production by photocatalytic methods is the reactor design. Most laboratory-scale systems are relatively controllable under laboratory conditions and in small volumes. However, when these systems are scaled up for large-scale production, some significant problems arise. Therefore, photocatalysis-based methods must develop highly efficient and easily controllable reactors for large-scale H_2 production applications.

In photocatalytic processes, the materials used in reactor manufacture and the photocatalysts used in photocatalytic processes are expected to have properties such as chemical stability, long-term preservation of catalytic activity, and reusability over a wide range of pH values. They should be durable in harsh photocatalytic reactions and resistant to degradation over time. Developing materials that can withstand long-term use is essential for long-term scalability. Heterogeneous photocatalytic systems have a relatively simple structure and are more economical than panel photocatalysts or photoelectrochemical systems [31]. Nevertheless, heterogeneous photocatalytic systems based on particulate photocatalysts have limited use in large-scale applications [39–42]. These limitations mainly stem from three different issues:

(1) Heterogeneous systems require a lot of water. This makes the operation of these processes difficult and uneconomical.
(2) The settling of particles to the bottom of the vessel is a major problem. If this occurs, it decreases the photocatalyst's interaction with light, thereby decreasing the process efficiency. Therefore, continuous and effective mixing is required.
(3) Difficulties in separating and recovering the photocatalyst from the suspension at the end of the process [1].

When designing reactors to be used for heterogeneous photocatalytic systems, operational parameters such as the water flow rate, catalyst dosage, and pollutant concentration, which potentially have direct effects on the efficiency of the photocatalytic process, must be taken into account [31, 39].

7.4 Advances in photocatalytic materials and innovative synthesis approaches

'Photocatalytic materials' is a general term for substances that have the potential to initiate or accelerate chemical reactions when exposed to light. They are widely used in applications such as water splitting for environmental remediation and hydrogen production. The development of efficient and cost-effective photocatalysts is an exciting area of research.

Titanium dioxide (TiO_2) and zinc oxide (ZnO) are the best-known photocatalysts; these materials are also called conventional photocatalysts. Numerous scientific studies have been conducted on the photocatalytic properties of these materials, and their stability and photocatalytic efficiency in the degradation of many pollutants and water splitting processes have been repeatedly proven. To increase the efficiency of these photocatalysts, composite or hybrid forms with other materials have also been produced. However, researchers continue working with these materials to improve efficiency and find new applications. In this context, hybrid catalysts produced by doping conventional photocatalysts such as titanium dioxide (TiO_2) with metals or nonmetals are known to improve charge separation and extend light absorption into the visible range, which can lead to increased photocatalytic efficiency. Therefore, it is expected that many scientific studies will be carried out in the future to develop new TiO_2-based hybrid or composite photocatalysts [43].

MOFs are a group of materials that have recently been extensively used for different purposes in many other applications. MOFs are composed of metal ions coordinated with organic ligands. These materials offer great potential for photocatalytic applications due to their high surface area and tunable properties. Studies are being carried out to improve the light absorption properties of MOFs. This is expected to further enhance the photocatalytic properties of these materials. MOFs can also be combined with other materials used as catalysts, significantly increasing their efficiency [44–47].

Perovskites are another group of materials whose photocatalytic properties have been studied in detail. Perovskite materials, which stand out due to their high potential for use in solar cell technology, have also been extensively investigated for their efficiency in photocatalytic processes. Perovskites also have great potential as photocatalysts due to their high photosensitivity. The high photosensitivity of perovskite materials allows them to absorb light emitted over an extensive wavelength range. Therefore, they are potentially efficient photocatalysts [48–50].

Quantum dots are another group of materials used as photocatalysts today. Quantum dots are semiconducting nanocrystals with a tunable bandgap structure. They can be incorporated into the structure of various photocatalysts to improve light efficiency and charge separation. Thanks to their flexible and tunable bandgap structure, these materials enable precise control of the light absorption properties of the materials they are incorporated into. Thus, they can make it possible to prepare highly efficient photocatalysts [51, 52].

Two-dimensional (2D) materials such as graphene, graphene oxide (GO), and transition-metal dichalcogenides (MXenes) can exhibit photocatalytic effects themselves or can be used to enhance the activity of other photocatalysts. These materials are widely used in photocatalysis-based applications due to their high surface area and advanced electronic and optical properties.

Graphene, GO, and reduced GO are carbon-based materials with different structures. These materials are widely used in photocatalytic processes for various purposes. Pure graphene has limited photocatalytic activity due to its zero bandwidth. Its zero bandgap means that graphene cannot absorb visible light efficiently. However, it can exhibit a little photocatalytic activity under UV light. Graphene is often used as a support material in photocatalytic systems to improve the efficiency of other photocatalysts [53, 54]. Its high surface area and excellent electron transport properties enhance the interaction between pollutants and the catalyst material. GO has a broader bandgap range than that of pure graphene. As a result, it can absorb visible light. GO can be used directly in photocatalytic processes or as a support material or cocatalyst for some photocatalysts. It can exhibit photocatalytic activity, especially for the degradation of some organic pollutants. GO is used in environmental remediation applications such as water treatment, pollutant degradation, and organic removal. When combined with other photocatalysts, it has the potential to be used in more advanced oxidation processes. Reduced graphene oxide (rGO) is obtained by the chemical reduction of GO. It has much improved photocatalytic activity compared to those of graphene and GO. The reduction process reverts GO to a graphene-like structure, reducing its bandgap. The reduction in bandgap allows rGO to absorb and utilize visible light more efficiently. rGO finds use in environmental remediation applications such as the degradation of organic pollutants, hydrogen production, and CO_2 reduction. Its enhanced photocatalytic properties make it a valuable material for various photocatalytic systems [55, 56].

In summary, graphene has limited photocatalytic activity but has the potential to be used as a support or cocatalyst. GO has a relatively wider bandgap and can thus exhibit photocatalytic activity under visible light. rGO, obtained by chemical reduction of GO, has the potential to exhibit even better photocatalytic activity. These materials have significant potential in various photocatalytic processes, especially in environmental remediation and sustainable energy production. It is expected that more scientific studies of the use of these materials in photocatalytic processes will be carried out to increase or optimize their efficiency in photocatalytic processes.

Carbon nitride (C_3N_4) is another carbon-based material extensively used in photocatalytic processes. Carbon nitride is a polymeric material with a graphitic structure that exhibits photocatalytic activity in visible light. Like other carbon-derived materials, carbon nitrides are seen as promising, especially for applications such as photocatalytic water splitting and the degradation of organic pollutants [57–59].

MXenes are another material group that has recently attracted great interest due to their high activity in photocatalytic processes. MXenes are a class of 2D materials that have recently attracted much attention because of their unique structural, morphological, and electrical properties. MXenes include transition-metal carbides, nitrides, and carbonitrides. As a result of their tunable surfactant groups and high surface areas, MXenes are widely used in many environmental remediation applications. It is known that MXenes can be used as adsorbents, photocatalysts, sensors, and membranes in these environmental remediation processes [60, 61].

In many studies, MXene-based materials have been used as photocatalysts in some photocatalytic processes, exhibiting excellent photocatalytic activity. Most MXenes have a wide bandgap, one of the most fundamental properties for photocatalytic activity. As such, MXenes are materials with high potential for use in photocatalysis. MXenes and MXene-based composites show promise in environmental applications such as removing pollutants from water and air. Moreover, some MXenes and their composites have been extensively investigated for their potential use in hydrogen production via photocatalytic water splitting. In studies conducted for this purpose, it has been reported that they can split water into hydrogen and oxygen using sunlight and thus have the potential to constitute a sustainable approach to producing hydrogen as a clean energy source. Moreover, another critical feature of MXene-class materials is their ease of functionalization. It is estimated that the photocatalytic properties of MXenes can be further improved through various surface modifications or by combining them with other materials such as semiconductors or noble metal nanoparticles. Although MXenes are considered a promising class of materials for the future, there are still challenges in increasing their photocatalytic efficiency and optimizing their stability. Researchers continue to work actively to improve the performance of MXenes and to understand the mechanism of MXene-based photocatalytic processes [61, 62].

In summary, MXenes have exhibited promising photocatalytic properties in photocatalytic processes such as environmental remediation and hydrogen production by water splitting. In the coming years, MXenes or MXene-based materials will likely be used more intensively in photocatalytic processes, so the photocatalytic properties of MXenes will be better understood and utilized more efficiently.

Plasmonic photocatalysis is another basic approach recently applied to increase the efficiency of photocatalysts. It is known that plasmonic nanoparticles, such as gold and silver nanoparticles, can enhance photocatalytic activity through localized surface plasmon resonance [63]. When combined with a photocatalyst, plasmonic nanoparticles are known to significantly improve light absorption and charge separation in photocatalytic systems and thus notably increase the catalytic activity of photocatalysts [64–66].

Hybrid, composite, or heterojunction materials that combine different photocatalyst materials can also exhibit effective photocatalytic activities. A common approach to obtaining hybrid photocatalysts is to incorporate other materials by self-assembly. For example, it is known that when metallic nanoparticles are combined with semiconductors, the resulting photocatalyst can facilitate charge transfer and improve photocatalytic performance. Therefore, many different photocatalysts are now obtained by combining various materials.

Photocatalytic processes are undergoing rapid change and development. Researchers are constantly investigating different photocatalytic materials and synthesis methods to develop highly efficient and more functional photocatalysts.

To synthesize the materials used as photocatalysts, bottom-up approaches such as the sol–gel method, hydrothermal synthesis, and chemical vapor deposition are generally preferred, allowing precise control over the structures and properties of the product materials. Through these methods, the morphological properties of the produced catalysts, such as shape and size (which have the potential to directly affect their catalytic activity), can also be precisely controlled. Bottom-up synthesis methods are expected to be extensively used in the synthesis of new photocatalysts that are likely to emerge in the future.

Another synthesis approach that is expected to be widely used in the production of photocatalysts is bioinspired synthesis. Bioinspired synthesis is a synthesis approach based on the creation of new materials inspired by natural photosynthetic systems. It is known that many photochemical processes occur spontaneously in nature [67]. Researchers aim to design new materials and processes inspired by these natural processes and to produce new photocatalysts. Moreover, the development of artificial systems that simulate natural photosynthesis is expected to become an essential application of photocatalysis in the coming years. These systems will convert solar energy into chemical fuels such as hydrogen or directly into valuable products [68].

The synthesis and characterization of nanostructured materials with high surface area and well-defined structures is an important research area. Nanomaterials can be produced by controlling their size, shape, and chemical composition. It is known that the photocatalytic properties of nanomaterials, like many other properties, are significantly affected by morphological properties such as their shape and size. Therefore, it may be possible to tailor nanomaterials for specific reactions. Like all nanotechnology-based products and processes, nanomaterials' production and synthesis methods are a constantly evolving research field. As a result of new technologies, it is possible to design and produce nanostructures with special photocatalytic properties is possible. In the future, it is expected that advanced synthesis methods such as atomic-layer deposition and chemical vapor deposition will be used more intensively in synthesizing photocatalyst materials. Thus, more effective new photocatalysts will be developed.

Machine learning and computational chemistry approaches are other synthesis approaches currently attracting interest for use in the synthesis of new materials. As a result of these methods, it is possible to predict the properties and optimize the structures of new photocatalysts that are intended to be synthesized before proceeding to the experimental stage. Machine learning and computational chemistry approaches are expected to be used more intensively in the synthesis of photocatalysts. Thus, the objective is to design more advanced photocatalysts and, at the same time, to save time and resources in the laboratory [69–72].

Another synthesis approach used extensively for designing more advanced photocatalysts is cocrystallization. This method is based on developing composite materials by cocrystallizing different chemical compounds. Photocatalysts produced by this method are expected to exhibit synergistic effects that enhance photocatalytic performance in addition to the individual photocatalytic activity of the materials [73, 74].

Another synthesis approach that has attracted interest in nanomaterials is *in situ* synthesis. Today, most of the materials used as photocatalysts are nanomaterials. All methods for producing nanomaterials can be categorized into two main groups: top-down and bottom-up synthesis. Top-down synthesis methods are generally more straightforward and better suited to being scaled up. However, the nanomaterials produced by this method are relatively large. They are also inhomogeneous in shape and size. Moreover, the yield is low, and the synthesis process offers limited control [28, 29, 75, 76]. However, bottom-up methods involving highly complex chemical processes make it possible to synthesize small and homogeneous nanoparticles [77, 78]. Bottom-up synthesis methods can be carried out in a single step (one-pot) or multiple stages. Among the bottom-up approaches, one-pot [79]and *in situ* synthesis methods are the most suitable for the large-scale, low-cost production of nanomaterials [80]. In particular, one-pot synthesis methods are considered economical and environmentally friendly, as they involve few reaction steps [81, 82]. The *in situ* synthesis method is a synthesis approach that aims to synthesize nanomaterials directly in the application environment, thereby providing photocatalysts with enhanced integration capability and efficiency [83, 84]. Moreover, the use of *in situ* synthesis methods makes it possible to precisely control the morphological properties of the product material, such as its shape and size [85, 86]. In the coming years, *in situ* synthesis approaches are expected to be applied more intensively and effectively to the synthesis of new photocatalysts.

It is possible to obtain photocatalysts under laboratory conditions, even if the synthesis processes are complex. However, large-scale synthesis processes require large reactors, and the methods of ensuring homogeneous and high-temperature heating and precise control of mass transfer in these systems can be much more complex [87]. Moreover, large-scale nanomaterial production may require several methods to be combined [88]. Furthermore, possible by-products or wastes may be generated during these processes. The potential adverse environmental impacts of production processes need to be carefully considered [85, 89]

The future of photocatalyst materials and their synthesis methods will likely be shaped by combining these approaches, driven by the need for more efficient and sustainable catalysts for various applications, including renewable energy production and environmental remediation. It is crucial to design advanced photocatalysts to address the global challenges of clean energy and pollution control.

7.5 The future evolution of light sources for photocatalytic processes

The choice of a light source for a photocatalytic process is a crucial parameter, as it directly affects the efficiency and applicability of the process. The chosen light sources for future photocatalysis processes are likely to be significantly influenced by several factors, such as energy efficiency, spectral range, and cost-effectiveness. Some potential trends related to the light sources used in future photocatalytic processes are summarized below.

LEDs are light sources that have high energy efficiency. Catalysts with different chemical structures are known to be activated when exposed to different wavelengths of radiation. LED light sources have the advantage that they can be tuned to emit specific wavelengths to optimize the photocatalytic activity of other materials [90–92]. Therefore, they are likely to be widely used as light sources, especially in applications requiring variable or precise light spectrum control.

Xenon and mercury lamps emit light over a wider spectral range, including the UV and visible-light regions. These lamps, which emit continuous and stable radiation, have become the traditional light sources used for photocatalytic processes. Although they are less energy efficient than LEDs in terms of the light intensity they emit, they are widely preferred when a wide spectral range is required. Therefore, xenon and mercury lamps are expected to be extensively used as light sources for photocatalytic processes in the future.

One of the approaches that has recently attracted interest in the field of photocatalysis is the use of plasmonic nanoparticles as light sources. It is expected that much more extensive research will be carried out on photocatalytic technology based on plasmonic particles, which is still under development. Plasmonic nanoparticles have the potential to significantly increase the absorption and utilization efficiency of light in photocatalytic reactions. Therefore, gold and silver nanoparticles, known for their strong plasmonic properties, have the potential to play an essential role in increasing the efficiency of photocatalytic processes.

The light source is a factor that directly impacts the sustainability of photocatalytic processes. Obtaining the light required for the process from natural sunlight is a sustainable and cost-effective approach. Therefore, photocatalytic processes that efficiently utilize the sunlight spectrum will likely become more prominent in the near term. On the other hand, technologies that allow more concentrated sunlight to be used are developing rapidly. Concentrated solar energy systems have the potential to provide intense, stable, and continuous light for specific applications. Therefore, much work is expected to be done on integrating concentrated solar energy systems into photocatalytic processes. Pulsed light sources such as pulsed xenon lamps and lasers can be used in specific photocatalytic applications in which intense, short bursts of light are required. They can potentially be used for fast, high-energy reactions [91, 93].

Future photocatalytic processes may also require the use of hybrid light sources based on combining natural sunlight with artificial light to optimize energy efficiency. In this way, the adverse effects of variations in sunlight, which can vary continuously throughout the day, on the ongoing photocatalytic process can be significantly minimized.

As in almost all fields, the general trend when choosing light sources for photocatalytic processes is to use environmentally friendly and nonhazardous light sources. The cost of light sources is pivotal to the commercial viability of photocatalytic processes. Their price is likely to be reduced over time due to advances in manufacturing and technology. The choice of light source for a photocatalytic process varies depending on the application's requirements, including the photocatalyst's spectral absorption characteristics and the desired reaction kinetics. Thus, some applications, especially in sectors such as water treatment and air quality improvement, may require the development of specialized light sources tailored to specific catalysts and reactions. Like many other fields, light technology is constantly evolving and being renewed by advancing technology. Therefore, in the future, in parallel with advances in technology and materials science, it is expected that new, more efficient, cost-effective, and sustainable light sources will be developed and integrated into photocatalytic systems. Moreover, if optical systems based on the guidance or management of light, such as lenses, mirrors, and light guides, are integrated into photocatalytic systems, the efficiency of these processes will increase significantly. Thus, more sustainable strategies can be designed.

Studies of hydrogen production by the photocatalytic splitting of water or photocatalysis-based environmental remediation processes are usually carried out on an experimental scale in small-volume glass or quart reactors. In these systems, also known as heterogeneous photocatalysis processes, the entire active surface of the photocatalyst can be used. Thus, the photocatalysts used are highly effective, and these experimental-scale studies observe excellent yields and high reaction rates. However, due to the nature of heterogeneous mixtures for large-scale applications, it is also possible that the particles used as photocatalysts may settle to the bottom of the reactor, and thus, their effectiveness may decrease. To avoid this, it is essential to ensure that adequate mixing is performed from the beginning to the end of the photocatalytic process. This mixing can lead to extra energy consumption and additional costs in large-volume systems. As the efficiency of the photocatalyst decreases over time, the photocatalyst needs to be regenerated or replaced. However, it is a challenging process to separate these photocatalysts, which are primarily nanomaterials, from the water environment and collect them for reuse. Although methods such as magnetic separation, centrifugation, etc. can be used in laboratory-scale studies, they would need to be more effective for production-scale applications. Therefore, it is anticipated that photocatalysts immobilized on a fixed substrate may be more helpful and significantly increase the usability of photocatalytic processes in large-scale applications. The main expectations here are to facilitate the separation of spent photocatalysts from the aqueous solution and to provide continuous system operability. However, these systems have limiting factors, such as their inability to utilize the entire surface of the catalyst and the reduction of the irradiated catalyst surface, which reduce their efficiency [94–96].

Plasmonic photocatalysis can also be used in the photocatalytic degradation of organic pollutants in water and air. Plasmonic materials can improve the performance of solar cells and other energy conversion technologies by enhancing light absorption and charge separation. Moreover, plasmonic nanoparticles are used to develop sensitive sensors that can detect small concentrations of various substances.

Plasmonic photocatalysis has excellent potential to improve the efficiency of photocatalytic reactions. However, plasmonic photocatalysis is still an emerging research field and poses several challenges, including optimizing plasmonic nanomaterials, controlling plasmon resonance, and developing practical and scalable applications. Many scientific studies are expected to overcome these challenges in the future. Thus, it is predicted that this process will find more widespread applications.

7.5.1 Plasmonic photocatalysis

Plasmonic photocatalysis is a field of research based on combining particles with plasmonic properties and photocatalytic materials to improve the efficiency of photocatalytic reactions. The plasmonic effect occurs when electromagnetic radiation interacts with free electrons in the structure of a noble metal, such as gold or silver, resulting in the formation of surface plasmons on the metal surface. These surface plasmons have the potential to significantly enhance light absorption and energy transfer in photocatalytic processes. Plasmonic photocatalytic technology is based on the use of plasmonic particles, such as gold or silver nanoparticles, in combination with photocatalysts. Plasmonic nanoparticles exhibit a unique property known as surface plasmon resonance (SPR), which allows materials to absorb and re-scatter specific wavelengths of light incident on them. Plasmonic photocatalysis technology often combines plasmonic nanoparticles with conventional photocatalysts such as titanium dioxide or zinc oxide to form hybrid materials. These hybrid photocatalysts can utilize absorbed light to initiate photochemical reactions [66].

The mechanism of plasmonic photocatalysis takes place in several basic steps. In the first step, plasmonic nanoparticles absorb incident light due to SPR, forming localized surface plasmons. The energy of these plasmons can be tuned to match the desired wavelength, which is usually in the visible or near-infrared range. In the second stage, light interacts with the plasmonic nanoparticles, generating intense and localized electric fields around the plasmonic metal nanoparticles. These localized electric fields concentrate the light energy and initiate the photochemical process, resulting in the formation of electron (e^-) and hole (h+) pairs in the internal structure of the photocatalyst. In the final stage, the enhanced charge transfer and high energy concentration resulting from the plasmonic effect increase reaction rates and make the photocatalytic process more efficient [63, 65].

One of the potential application areas of plasmonic photocatalysis is photocatalytic water splitting processes. Plasmonic photocatalysts significantly increase the efficiency of reactions that produce hydrogen and oxygen gases by splitting water [66]. Plasmonic photocatalysis can also be used in the photocatalytic degradation of organic pollutants in water and air. Plasmonic materials can improve the performance of solar cells and other energy conversion technologies by enhancing light absorption and charge separation. Moreover, plasmonic nanoparticles are used to develop sensitive sensors that can detect small concentrations of various substances.

Plasmonic photocatalysis has excellent potential to improve the efficiency of photocatalytic reactions. However, plasmonic photocatalysis is still an emerging research field and poses several challenges, including optimizing plasmonic nanomaterials, controlling plasmon resonance, and developing practical and scalable applications. Many scientific studies are expected to overcome these challenges in the future. Thus, it is predicted that this process will find more widespread applications.

7.5.2 A comparison of visible-light photocatalysis and plasmonic photocatalysis

Visible-light photocatalysis and plasmonic photocatalysis differ significantly in terms of the light source used and the mechanism involved. In visible-light photocatalysis, visible light is the energy source. This makes it compatible with natural sunlight and typical indoor lighting. Visible-light photocatalysis uses photocatalytic materials to absorb visible light to form electron–hole pairs. These photoexcited charge carriers then reach the surface of the photocatalyst, where they participate in various chemical reactions. The active chemical species resulting from these reactions lead to the degradation of pollutants. A schematic representation of the visible-light photocatalytic mechanism is shown in figure 7.1. Visible-light photocatalysis is used in multiple applications, including water purification, air quality improvement, and organic synthesis.

In plasmonic photocatalysis processes, plasmonic nanoparticles such as gold or silver nanoparticles generate localized surface plasmons. These nanoparticles can absorb light at specific resonant wavelengths, usually in the visible or near-infrared range. Plasmonic photocatalysis requires the use of plasmonic nanoparticles in combination with a conventional photocatalyst. Plasmonic nanoparticles absorb light and form localized surface plasmons, generating intense and highly localized electric fields. These electric fields concentrate light energy near the active sites of the photocatalyst, promoting electron–hole pair formation and facilitating charge transfer. Thus, they increase reaction rates. Plasmonic photocatalysis has applications such as water splitting, pollutant degradation, and energy conversion. It is beneficial when visible or near-infrared light needs to be used efficiently or when visible light is insufficient for the photocatalytic reaction.

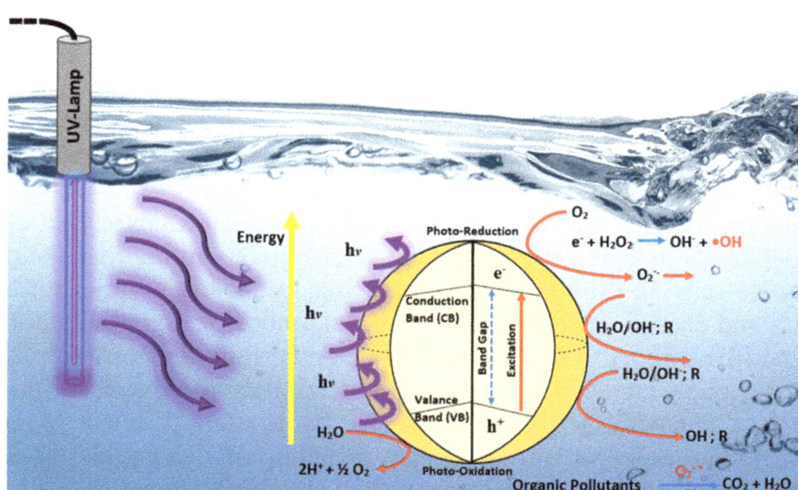

Figure 7.1. A schematic representation of the visible-light photocatalysis mechanism. Adapted from [97], Copyright (2022), with permission from Elsevier.

The main difference between visible-light photocatalysis and plasmonic photocatalysis processes is the light source used. Visible-light photocatalysis utilizes the visible light used in natural and indoor lighting. Plasmonic photocatalysis, on the other hand, uses plasmonic nanoparticles to concentrate light at specific wavelengths across a wide range of the electromagnetic wave spectrum, including the visible or near-infrared regions. Another critical difference between these two processes is that they differ in their mechanisms. Plasmonic photocatalysis relies on generating localized surface plasmons and enhancing local electric fields to improve reaction rates. In contrast, visible-light photocatalysis uses visible light directly to excite charge carriers in the photocatalyst. On the other hand, both methods rely on the use of photocatalysts. However, plasmonic photocatalytic processes require the use of plasmonic nanoparticles in addition to photocatalysts. When a choice between these methods has to be made, an assessment should be performed depending on the specific requirements of the photocatalytic reaction and the available light source.

7.6 Conclusions

Currently, only a few photocatalysts can be used in large-scale photocatalytic applications involving environmental applications or water splitting. The main reason for the limited availability of commercially available photocatalysts is that many new photocatalysts developed need to meet essential criteria, such as low cost, high efficiency, chemical stability, stability, and environmental friendliness. Therefore, photocatalyst production processes must be carefully considered, and the operating parameters must be optimized for commercial-scale production.

The techniques proposed for the synthesis of photocatalysts in recent years have different advantages and weaknesses in terms of catalyst stability, efficiency, cost, production, structure, and catalytic performance. Many catalyst materials and production methods for these materials have been proposed in the studies carried out. However, most of the proposed synthesis techniques have various difficulties, such as uncontrollable catalyst morphology, a high risk of contamination, complex purification processes, labor-intensive methods, multistep processes, expensive starting chemicals, waste generation, environmental unfriendliness, low yield, and high energy consumption. Moreover, for a photocatalyst, critical factors such as the bandgap, the types of charge carriers, the crystallinity of the materials, the specific surface area, and the stability of the photocatalyst are all closely related to the synthesis technique used to produce the catalyst. All these factors potentially affect the electron–hole recombination process, the synergy between the pollutant and the photocatalyst, the availability of active sites, and the reusability of the photocatalyst. Considering all these aspects, it is predicted that the obstacles, limitations, and challenges encountered in developing new photocatalysts will continue to be an attractive research area for researchers for some time.

In summary, photocatalytic processes have great potential in many application areas, such as environmental sustainability, energy production, water treatment, and chemical synthesis. However, some critical challenges must be overcome to secure the wider adoption and development of this technology.

There is a need to improve the efficiency of materials used in photocatalytic processes and to design and manufacture more advanced materials capable of absorbing light over a wider electromagnetic wavelength range. Similarly, more work needs to be done on the design and efficiency of photocatalytic reactors. Parameters such as scalability, energy efficiency, and operating costs should also be considered when designing reactors. On the other hand, although many scientific studies have been conducted to identify the mechanisms of photocatalytic processes, these processes still need to be fully explained. Artificial intelligence may be an essential tool in understanding the mechanisms of photocatalysis and designing new materials. Therefore, more qualified studies on these topics will be needed. If the challenges of photocatalytic processes can be overcome, they have great potential for use in many applications, such as wastewater treatment, converting sunlight into electrical energy, controlling air pollution, and sustainable energy production by water splitting.

In conclusion, photocatalysis offers significant benefits in various application areas, but more research and development work needs to be done before it can be adapted to large-scale systems. Although the future of this technology is bright, a sustained effort is required to overcome the scientific and technical challenges. The wider adoption of photocatalysis would be an excellent contribution to a sustainable future.

References

[1] Kuspanov Z, Bakbolat B, Baimenov A, Issadykov A, Yeleuov M and Daulbayev C 2023 Photocatalysts for a sustainable future: innovations in large-scale environmental and energy applications *Sci. Total Environ.* **885** 163914

[2] Wang Q, Gao Q, Al-Enizi A M, Nafady A and Ma S 2020 Recent advances in MOF-based photocatalysis: environmental remediation under visible light *Inorg Chem. Front.* **7** 300–39

[3] Koe W S, Lee J W, Chong W C, Pang Y L and Sim L C 2020 An overview of photocatalytic degradation: photocatalysts, mechanisms, and development of photocatalytic membrane *Environ. Sci. Pollut. Res. Int.* **27** 2522–65

[4] Hoang S and Gao P X 2016 Nanowire array structures for photocatalytic energy conversion and utilization: a review of design concepts, assembly and integration, and function enabling *Adv. Energy Mater.* **6** 1600683

[5] Borges M E, Sierra M, Cuevas E, García R and Esparza P 2016 Photocatalysis with solar energy: sunlight-responsive photocatalyst based on TiO_2 loaded on a natural material for wastewater treatment *Sol. Energy* **135** 527–35

[6] Sun J-H, Dong S-Y, Feng J-L, Yin X-J and Zhao X-C 2011 Enhanced sunlight photocatalytic performance of Sn-doped ZnO for methylene blue degradation *J. Mol. Catal. A: Chem.* **335** 145–50

[7] Cai T, Liu Y, Wang L, Zhang S, Zeng Y, Yuan J, Ma J, Dong W, Liu C and Luo S 2017 Silver phosphate-based Z-Scheme photocatalytic system with superior sunlight photocatalytic activities and anti-photocorrosion performance *Appl. Catal.* B **208** 1–13

[8] Chen P, Blaney L, Cagnetta G, Huang J, Wang B, Wang Y, Deng S and Yu G 2019 Degradation of ofloxacin by perylene diimide supramolecular nanofiber sunlight-driven photocatalysis *Environ. Sci. Technol.* **53** 1564–75

[9] Enesca A and Isac L 2020 The influence of light irradiation on the photocatalytic degradation of organic pollutants *Materials (Basel)* **13** 2494
[10] Maeda K 2011 Photocatalytic water splitting using semiconductor particles: history and recent developments *J. Photochem. Photobiol.,* C **12** 237–68
[11] Jafari T, Moharreri E, Amin A S, Miao R, Song W and Suib S L 2016 Photocatalytic water splitting—the untamed dream: a review of recent advances *Molecules* **21** 900
[12] Maeda K and Domen K 2010 Photocatalytic water splitting: recent progress and future challenges *J. Phys. Chem. Lett.* **1** 2655–61
[13] Gupta A, Likozar B, Jana R, Chanu W C and Singh M K 2022 A review of hydrogen production processes by photocatalytic water splitting—from atomistic catalysis design to optimal reactor engineering *Int. J. Hydrogen Energy* **47** 33282–307
[14] Goto Y, Hisatomi T, Wang Q, Higashi T, Ishikiriyama K, Maeda T, Sakata Y, Okunaka S, Tokudome H and Katayama M 2018 A particulate photocatalyst water-splitting panel for large-scale solar hydrogen generation *Joule* **2** 509–20
[15] Schneider J, Matsuoka M, Takeuchi M, Zhang J, Horiuchi Y, Anpo M and Bahnemann D W 2014 Understanding TiO_2 photocatalysis: mechanisms and materials *Chem. Rev.* **114** 9919–86
[16] Xu C, Anusuyadevi P R, Aymonier C, Luque R and Marre S 2019 Nanostructured materials for photocatalysis *Chem. Soc. Rev.* **48** 3868–902
[17] Yang X and Wang D 2018 Photocatalysis: from fundamental principles to materials and applications *ACS Appl. Energy Mater.* **1** 6657–93
[18] Luo B, Liu G and Wang L 2016 Recent advances in 2D materials for photocatalysis *Nanoscale* **8** 6904–20
[19] Gholipour M R, Dinh C-T, Béland F and Do T-O 2015 Nanocomposite heterojunctions as sunlight-driven photocatalysts for hydrogen production from water splitting *Nanoscale* **7** 8187–208
[20] Idriss H 2020 The elusive photocatalytic water splitting reaction using sunlight on suspended nanoparticles: is there a way forward? *Catal. Sci. Technol.* **10** 304–10
[21] Kumar P, Boukherroub R and Shankar K 2018 Sunlight-driven water-splitting using two-dimensional carbon based semiconductors *J. Mater. Chem.* A **6** 12876–931
[22] Tentu R D and Basu S 2017 Photocatalytic water splitting for hydrogen production *Curr. Opin. Electrochem.* **5** 56–62
[23] Braham R J and Harris A T 2009 Review of major design and scale-up considerations for solar photocatalytic reactors *Ind. Eng. Chem. Res.* **48** 8890–905
[24] Ray A K 1999 Design, modelling and experimentation of a new large-scale photocatalytic reactor for water treatment *Chem. Eng. Sci.* **54** 3113–25
[25] Alalm M G, Djellabi R, Meroni D, Pirola C, Bianchi C L and Boffito D C 2021 Toward scaling-up photocatalytic process for multiphase environmental applications *Catalysts* **11** 562
[26] Reilly K, Wilkinson D P and Taghipour F 2018 Photocatalytic water splitting in a fluidized bed system: computational modeling and experimental studies *Appl. Energy* **222** 423–36
[27] Abdel-Maksoud Y, Imam E and Ramadan A 2016 TiO_2 solar photocatalytic reactor systems: selection of reactor design for scale-up and commercialization—analytical review *Catalysts* **6** 138
[28] Burke D W, Sun C, Castano I, Flanders N C, Evans A M, Vitaku E, McLeod D C, Lambeth R H, Chen L X and Gianneschi N C 2020 Acid exfoliation of imine-linked covalent organic frameworks enables solution processing into crystalline thin films *Angew. Chem.* **132** 5203–9

[29] Chen X, Li Y, Wang L, Xu Y, Nie A, Li Q, Wu F, Sun W, Zhang X and Vajtai R 2019 High-lithium-affinity chemically exfoliated 2D covalent organic frameworks *Adv. Mater.* **31** 1901640

[30] Frowijn L S and van Sark W G 2021 Analysis of photon-driven solar-to-hydrogen production methods in the Netherlands *Sustain. Energy Technol. Assess.* **48** 101631

[31] Hisatomi T and Domen K 2019 Reaction systems for solar hydrogen production via water splitting with particulate semiconductor photocatalysts *Nat. Catal.* **2** 387–99

[32] Cao F, Wei Q, Liu H, Lu N, Zhao L and Guo L 2018 Development of the direct solar photocatalytic water splitting system for hydrogen production in Northwest China: design and evaluation of photoreactor *Renew. Energy* **121** 153–63

[33] Ruiz-Aguirre A, Villachica-Llamosas J, Polo-López M, Cabrera-Reina A, Colón G, Peral J and Malato S 2022 Assessment of pilot-plant scale solar photocatalytic hydrogen generation with multiple approaches: valorization, water decontamination and disinfection *Energy* **260** 125199

[34] Ren H, Koshy P, Chen W-F, Qi S and Sorrell C C 2017 Photocatalytic materials and technologies for air purification *J. Hazard. Mater.* **325** 340–66

[35] Zhao J and Yang X 2003 Photocatalytic oxidation for indoor air purification: a literature review *Build. Environ.* **38** 645–54

[36] Hay S O, Obee T, Luo Z, Jiang T, Meng Y, He J, Murphy S C and Suib S 2015 The viability of photocatalysis for air purification *Molecules* **20** 1319–56

[37] He F, Jeon W and Choi W 2021 Photocatalytic air purification mimicking the self-cleaning process of the atmosphere *Nat. Commun.* **12** 2528

[38] Qian H, Hou Q, Zhang W, Nie Y, Lai R, Ren H, Yu G, Bai X, Wang H and Ju M 2022 Construction of electron transport channels and oxygen adsorption sites to modulate reactive oxygen species for photocatalytic selective oxidation of 5-hydroxymethylfurfural to 2, 5-diformylfuran *Appl. Catal.* B **319** 121907

[39] Jing D, Guo L, Zhao L, Zhang X, Liu H, Li M, Shen S, Liu G, Hu X and Zhang X 2010 Efficient solar hydrogen production by photocatalytic water splitting: from fundamental study to pilot demonstration *Int. J. Hydrogen Energy* **35** 7087–97

[40] Chen J, Xu W, Zuo H, Wu X, Jiaqiang E, Wang T, Zhang F and Lu N 2019 System development and environmental performance analysis of a solar-driven supercritical water gasification pilot plant for hydrogen production using life cycle assessment approach *Energy Convers. Manage.* **184** 60–73

[41] Chen S, Ma G, Wang Q, Sun S, Hisatomi T, Higashi T, Wang Z, Nakabayashi M, Shibata N and Pan Z 2019 Metal selenide photocatalysts for visible-light-driven Z-scheme pure water splitting *J. Mater. Chem.* A **7** 7415–22

[42] Chen X, Cai S, Yu E, Li J, Chen J and Jia H 2019 Photothermocatalytic performance of ACo_2O_4 type spinel with light-enhanced mobilizable active oxygen species for toluene oxidation *Appl. Surf. Sci.* **484** 479–88

[43] Kubiak A 2023 Comparative study of TiO_2–Fe_3O_4 photocatalysts synthesized by conventional and microwave methods for metronidazole removal *Sci. Rep.* **13** 12075

[44] Li Y, Xu H, Ouyang S and Ye J 2016 Metal–organic frameworks for photocatalysis *Phys. Chem. Chem. Phys.* **18** 7563–72

[45] García-Salcido V, Mercado-Oliva P, Guzmán-Mar J L, Kharisov B I and Hinojosa-Reyes L 2022 MOF-based composites for visible-light-driven heterogeneous photocatalysis: synthesis, characterization and environmental application studies *J. Solid State Chem.* **307** 122801

[46] Navalón S, Dhakshinamoorthy A, Álvaro M, Ferrer B and García H 2023 Metal–organic frameworks as photocatalysts for solar-driven overall water splitting *Chem. Rev.* **123** 445–90
[47] Sun K, Qian Y and Jiang H-L 2023 Metal-organic frameworks for photocatalytic water splitting and CO_2 reduction *Angew. Chem. Int. Ed.* **62** e202217565
[48] DuBose J T and Kamat P V 2022 Efficacy of perovskite photocatalysis: challenges to overcome *ACS Energy Lett.* **7** 1994–2011
[49] Dandia A, Saini P, Sharma R and Parewa V 2020 Visible light driven perovskite-based photocatalysts: a new candidate for green organic synthesis by photochemical protocol *Curr. Res. Green Sustain. Chem.* **3** 100031
[50] Mohd Kaus N H, Ibrahim M L, Imam S S, Mashuri S I S and Kumar Y 2022 Efficient Visible-Light-Driven Perovskites Photocatalysis: Design, Modification and Application *Green Photocatalytic Semiconductors: Recent Advances and Applications* ed S Garg and A Chandra (Cham: Springer International Publishing) 357–98
[51] Sun P, Xing Z, Li Z and Zhou W 2023 Recent advances in quantum dots photocatalysts *Chem. Eng. J.* **458** 141399
[52] Chakraborty I N, Roy P and Pillai P P 2023 Visible light-mediated quantum dot photocatalysis enables olefination reactions at room temperature *ACS Catal.* **13** 7331–8
[53] Li X, Yu J, Wageh S, Al-Ghamdi A A and Xie J 2016 Graphene in photocatalysis: a review *Small* **12** 6640–96
[54] Lu K-Q, Li Y-H, Tang Z-R and Xu Y-J 2021 Roles of graphene oxide in heterogeneous photocatalysis *ACS Materials Au* **1** 37–54
[55] Albero J, Mateo D and García H 2019 Graphene-based materials as efficient photocatalysts for water splitting *Molecules* **24** 906
[56] Suresh R, Mangalaraja R V, Mansilla H D, Santander P and Yáñez J 2020 Reduced Graphene Oxide-Based Photocatalysis *Green Photocatalysts* ed M Naushad *et al* (Cham: Springer International Publishing) 145–66
[57] Singh P P and Srivastava V 2022 Recent advances in visible-light graphitic carbon nitride (g-C_3N_4) photocatalysts for chemical transformations *RSC Adv.* **12** 18245–65
[58] Gao R-H, Ge Q, Jiang N, Cong H, Liu M and Zhang Y-Q 2022 Graphitic carbon nitride (g-C_3N_4)-based photocatalytic materials for hydrogen evolution *Front. Chem.* **10**
[59] Ong W-J, Tan L-L, Ng Y H, Yong S-T and Chai S-P 2016 Graphitic carbon nitride (g-C_3N_4)-based photocatalysts for artificial photosynthesis and environmental remediation: are we a step closer to achieving sustainability? *Chem. Rev.* **116** 7159–329
[60] Gogotsi Y and Anasori B 2019 *The rise of MXenes* (Washington, DC: ACS Publications) 8491–4
[61] Kuang P, Low J, Cheng B, Yu J and Fan J 2020 MXene-based photocatalysts *Journal of Materials Science & Technology* **56** 18–44
[62] Zhong Q, Li Y and Zhang G 2021 Two-dimensional MXene-based and MXene-derived photocatalysts: recent developments and perspectives *Chem. Eng. J.* **409** 128099
[63] Zhang X, Chen Y L, Liu R-S and Tsai D P 2013 Plasmonic photocatalysis *Rep. Prog. Phys.* **76** 046401
[64] Verma R, Belgamwar R and Polshettiwar V 2021 Plasmonic photocatalysis for CO_2 conversion to chemicals and fuels *ACS Mater. Lett.* **3** 574–98
[65] Wang T, Wang H-J, Lin J-S, Yang J-L, Zhang F-L, Lin X-M, Zhang Y-J, Jin S and Li J-F 2023 Plasmonic photocatalysis: mechanism, applications and perspectives *Chinese J. Struct. Chem.* **42** 100066

[66] Díaz F J P, del Río R S and Rodriguez P E D S 2022 Plasmonic Photocatalysts for Water Splitting *Photoelectrochemical Hydrogen Generation: Theory, Materials Advances, and Challenges* ed P Kumar and P Devi (Singapore: Springer Nature Singapore) 117–73

[67] Jiang X, Chen Y-X and Lu C-Z 2020 Bio-inspired materials for photocatalytic hydrogen production *Chinese J. Struct. Chem.* **39** 2123–30

[68] Fu J, Zhu B, You W, Jaroniec M and Yu J 2018 A flexible bio-inspired H_2-production photocatalyst *Appl. Catal.* B **220** 148–60

[69] Masood H, Toe C Y, Teoh W Y, Sethu V and Amal R 2019 Machine learning for accelerated discovery of solar photocatalysts *ACS Catal.* **9** 11774–87

[70] Mai H, Le T C, Chen D, Winkler D A and Caruso R A 2022 Machine learning for electrocatalyst and photocatalyst design and discovery *Chem. Rev.* **122** 13478–515

[71] Keith J A, Vassilev-Galindo V, Cheng B, Chmiela S, Gastegger M, Müller K-R and Tkatchenko A 2021 Combining machine learning and computational chemistry for predictive insights into chemical systems *Chem. Rev.* **121** 9816–72

[72] Li X, Maffettone P M, Che Y, Liu T, Chen L and Cooper A I 2021 Combining machine learning and high-throughput experimentation to discover photocatalytically active organic molecules *Chem. Sci.* **12** 10742–54

[73] Liu C, Huang H, Du X, Zhang T, Tian N, Guo Y and Zhang Y 2015 *In situ* co-crystallization for fabrication of g-C_3N_4/Bi_5O_7I heterojunction for enhanced visible-light photocatalysis *J. Phys. Chem.* C **119** 17156–65

[74] Ágoston Á and Janovák L 2023 Hydrothermal co-crystallization of novel copper tungstate-strontium titanate crystal composite for enhanced photocatalytic activity and increased electron–hole recombination time *Catalysts* **13** 287

[75] Li P, Yan X, He Z, Ji J, Hu J, Li G, Lian K and Zhang W 2016 α-Fe_2O_3 concave and hollow nanocrystals: top-down etching synthesis and their comparative photocatalytic activities *CrystEngComm.* **18** 1752–9

[76] Yusran Y, Li H, Guan X, Li D, Tang L, Xue M, Zhuang Z, Yan Y, Valtchev V and Qiu S 2020 Exfoliated mesoporous 2D covalent organic frameworks for high-rate electrochemical double-layer capacitors *Adv. Mater.* **32** 1907289

[77] Cui L, Liu Y, Fang X, Yin C, Li S, Sun D and Kang S 2018 Scalable and clean exfoliation of graphitic carbon nitride in NaClO solution: enriched surface active sites for enhanced photocatalytic H_2 evolution *Green Chem.* **20** 1354–61

[78] Liu W, Li X, Wang C, Pan H, Liu W, Wang K, Zeng Q, Wang R and Jiang J 2019 A scalable general synthetic approach toward ultrathin imine-linked two-dimensional covalent organic framework nanosheets for photocatalytic CO_2 reduction *J. Am. Chem. Soc.* **141** 17431–40

[79] Tian B, Tian B, Smith B, Scott M, Lei Q, Hua R, Tian Y and Liu Y 2018 Facile bottom-up synthesis of partially oxidized black phosphorus nanosheets as metal-free photocatalyst for hydrogen evolution *Proc. Natl. Acad. Sci.* **115** 4345–50

[80] Liu M, Xing Z, Li Z and Zhou W 2021 Recent advances in core–shell metal organic frame-based photocatalysts for solar energy conversion *Coord. Chem. Rev.* **446** 214123

[81] Chen J, Tang T, Feng W, Liu X, Yin Z, Zhang X, Chen J and Cao S 2021 Large-scale synthesis of p–n heterojunction Bi_2O_3/TiO_2 nanostructures as photocatalysts for removal of antibiotics under visible light *ACS Appl. Nano Mater.* **5** 1296–307

[82] Wang L, Hong Y, Liu E, Wang Z, Chen J, Yang S, Wang J, Lin X and Shi J 2020 Rapid polymerization synthesizing high-crystalline g-C_3N_4 towards boosting solar photocatalytic H_2 generation *Int. J. Hydrogen Energy* **45** 6425–36

[83] Weng S, Chen B, Xie L, Zheng Z and Liu P 2013 Facile *in situ* synthesis of a Bi/BiOCl nanocomposite with high photocatalytic activity *J. Mater. Chem.* A **1** 3068–75
[84] Ge L, Han C and Liu J 2012 *In situ* synthesis and enhanced visible light photocatalytic activities of novel PANI–gC_3N_4 composite photocatalysts *J. Mater. Chem.* **22** 11843–50
[85] Li G, Zhang K, Li C, Gao R, Cheng Y, Hou L and Wang Y 2019 Solvent-free method to encapsulate polyoxometalate into metal-organic frameworks as efficient and recyclable photocatalyst for harmful sulfamethazine degrading in water *Appl. Catalysis* B **245** 753–9
[86] Xin X, Song Y, Guo S, Zhang Y, Wang B, Yu J and Li X 2020 In-situ growth of high-content 1T phase MoS_2 confined in the CuS nanoframe for efficient photocatalytic hydrogen evolution *Appl. Catal.* B **269** 118773
[87] Yi J, Fei T, Li L, Yu Q, Zhang S, Song Y, Lian J, Zhu X, Deng J and Xu H 2021 Large-scale production of ultrathin carbon nitride-based photocatalysts for high-yield hydrogen evolution *Appl. Catalysis* B **281** 119475
[88] Deng S, Liu C, Zhang Y, Ji Y, Mei B, Yao Z and Lin S 2023 Large-scale preparation of ultrathin bimetallic nickel iron sulfides branch nanoflake arrays for enhanced hydrogen evolution reaction *Catalysts* **13** 174
[89] Jahanshahi R, Khazaee A, Sobhani S and Sansano J M 2020 g-C_3N_4/γ-Fe_2O_3/TiO_2/Pd: a new magnetically separable photocatalyst for visible-light-driven fluoride-free Hiyama and Suzuki–Miyaura cross-coupling reactions at room temperature *New J. Chem.* **44** 11513–26
[90] Izadifard M, Achari G and Langford C H 2013 Application of photocatalysts and LED light sources in drinking water treatment *Catalysts* **3** 726–43
[91] Jo W-K and Tayade R J 2014 New generation energy-efficient light source for photocatalysis: LEDs for environmental applications *Ind. Eng. Chem. Res.* **53** 2073–84
[92] Sergejevs A, Clarke C, Allsopp D, Marugan J, Jaroenworaluck A, Singhapong W, Manpetch P, Timmers R, Casado C and Bowen C 2017 A calibrated UV-LED based light source for water purification and characterisation of photocatalysis *Photochemical & Photobiological Sciences* **16** 1690–9
[93] Gondal M A, Ali M A, Chang X F, Shen K, Xu Q Y and Yamani Z H 2012 Pulsed laser-induced photocatalytic reduction of greenhouse gas CO_2 into methanol: a value-added hydrocarbon product over SiC *J. Environ. Sci. Health, Part* A **47** 1571–6
[94] Bui V K H, Tran V V, Moon J-Y, Park D and Lee Y-C 2020 Titanium dioxide microscale and macroscale structures: a mini-review *Nanomaterials* **10** 1190
[95] Loeb S K, Alvarez P J, Brame J A, Cates E L, Choi W, Crittenden J, Dionysiou D D, Li Q, Li-Puma G and Quan X 2019 The Technology Horizon for Photocatalytic Water Treatment: Sunrise or Sunset? *Environ. Sci. Technol.* **53** 2937–47
[96] Phan D D, Babick F, Trịnh T H T, Nguyen M T, Samhaber W and Stintz M 2018 Investigation of fixed-bed photocatalytic membrane reactors based on submerged ceramic membranes *Chem. Eng. Sci.* **191** 332–42
[97] Karakaş Z K 2022 A comprehensive study on the production and photocatalytic activity of copper ferrite nanoparticles synthesized by microwave-assisted combustion method as an effective photocatalyst *J. Phys. Chem. Solids* **170** 110927

www.ingramcontent.com/pod-product-compliance
Ingram Content Group UK Ltd.
Pitfield, Milton Keynes, MK11 3LW, UK
UKHW051853100926
8556IPUK00005B/607

9 780750 35